## Praise for *Driving the Future*

"The author has a vision for the future of the automobile. It's not exactly the flying car of the future, but almost, as it comes with smartphone-synced scheduling, zero-emissions technology, and the ability to park itself....Astute ... Oge knows her stuff." —*Kirkus Reviews*

"Margo tells the incredible story of how California and then Washington were able to mandate much cleaner cars and light trucks. This is the story of how hard it is to combat climate change—and also how imaginative and determined leaders can get it done." —**Governor Jerry Brown**

"Margo Oge describes the astounding transformation of cars and trucks in America—cutting pollution by more than 97 percent, and greenhouse gases by more than half—and shows the way to complete this job. It is a compelling story and a great read." —**Hal Harvey, CEO of Energy Innovation**

"*Driving the Future* is a testament to the progress that is possible when committed public servants are allowed to follow the scientific evidence where it leads and to envision and then execute ambitious plans for a better technological and environmental future."
— **Lisa Heinzerling, Justice William J. Brennan, Jr. Professor of Law, Georgetown University Law Center**

"Margo Oge provides a riveting insider's account of the people, science, politics, and technologies behind an improbable victory in the battle against global warming. Based on her understanding of how regulation can drive innovation, she depicts a future in which cleaner, lighter, smarter cars will become tools in the fight against climate change rather than contribute to it. Every citizen should read her book and feel proud of what we can accomplish together."
—**Lisa Jackson, EPA administrator under President Barack Obama**

"Drawing upon her experience as an architect of game-changing fuel economy standards, Margo Oge gives us a roadmap to a future world of better and cleaner cars, healthier air, reduced geopolitical conflict, and stronger communities. If we get there, it will be because of visionaries such as Oge."
— **Ken Kimmell, president, Union of Concerned Scientists**

"Margo and her team at the EPA helped craft far-reaching GHG and fuel economy standards for the US that are accelerating the adoption of future automotive technologies like plug-in electric vehicles and fuel cell vehicles. Margo has the credibility and the record for convincingly discussing a carbon free future in her book, a vision we share."
—**Ulrich Kranz, BMW, senior vice president, Product Line i**

"*Driving the Future* is a real-world story about the policy visionaries, business leaders, and dedicated citizens who are spurring the clean energy revolution—and ushering in a new era of prosperity for our nation and the world."
**—Fred Krupp, president, Environmental Defense Fund**

"Margo has given us a fascinating insider's view of regulatory development and a framework for collaborative rulemaking that literally changed the face—and tailpipes—of an industry, and a thoughtful look forward to future opportunities."
**—Tom Linebarger, chairman and CEO of Cummins Inc.**

"*Driving the Future* is the story of a dramatic success in a key battle in the fight against global warming—improving the environmental performance of vehicle fleets—with a lucid explanation of how to bridge science and public policy. Reading the book is as pleasant as having a talk with a good friend, and as informative as a full course in public policy."
**—Mario J. Molina, PhD, professor at the University of California, San Diego, and winner of the 1995 Nobel Prize in Chemistry for his role in elucidating the threat to the earth's ozone layer from chlorofluorocarbon gases**

"It's a great story: an insider's account of the unprecedented collaboration of politicians, regulators, and industry that created the world's first standards for low-carbon vehicles and a practical guide to steps that all of us—including consumers—must take to create a global market for vehicles that can take us where we want to go while dodging the worst effects of climate change."
**—Mary Nichols, chairman of the California Air Resources Board**

"Five stars to Margo Oge! She may be a veteran regulator, but she also has the vision thing, the broad historical and conceptual perspective that made her one of EPA's stars in the historic achievement of cleaner cars and in the cause of averting climate change. Not to mention that her writing is fluid, engaging, and makes a complex story hard to put down."
**—William Reilly, EPA administrator under President George H. W. Bush**

"This book should be a must-read for anyone who wants to know how the regulatory process actually works. Margo Oge provides a compelling insider's account about the making of President Obama's landmark vehicle emission rules. She describes how she and her dedicated team at EPA fought political interference under the Bush Administration, formed a crucial partnership with innovative California regulators, and found key allies within the new Obama team."
**—Congressman Henry Waxman**

# DRIVING THE FUTURE

# DRIVING THE FUTURE

## COMBATING CLIMATE CHANGE WITH CLEANER, SMARTER CARS

## MARGO T. OGE

**Arcade Publishing**
New York

First Edition

Arcade Publishing books may be purchased in bulk at special discounts for sales promotion, corporate gifts, fund-raising, or educational purposes. Special editions can also be created to specifications. For details, contact the Special Sales Department, Arcade Publishing, 307 West 36th Street, 11th Floor, New York, NY 10018 or arcade@skyhorsepublishing. com.

Arcade Publishing® is a registered trademark of Skyhorse Publishing, Inc.®, a Delaware corporation.

Visit our website at www.arcadepub.com.
Visit the author's website at margooge.com

10 9 8 7 6 5 4 3 2 1

Library of Congress Cataloging-in-Publication Data

Oge, Margo T.
Driving the future : combating climate change with cleaner, smarter cars / Margo
T. Oge. — First edition.
pages cm
Includes bibliographical references and index.
ISBN 978-1-62872-538-4 (hardcover : alk. paper); ISBN 978-1-62872-549-0 (ebook)
1. Alternative fuel vehicles.   2. Automobiles—Technological innovations.   3. Auto
mobiles—Environmental aspects.   4. Sustainable engineering.   I. Title.
TL216.5.O38 2015
629.222028'6—dc23     2014048036

Cover design and photograph by Enrique Florendo and Marisa Oge

Printed in the United States of America

To my husband, Cuneyt, my best friend, whose love, support, and belief in me means the world to me. To my daughters, Nicole and Marisa, who are the greatest gifts in my life. To my mother, Ioanna Tsirigotis, who is my hero.

To all EPA employees, especially my colleagues at the Office of Transportation and Air Quality, whose passion and hard work have resulted in significant improvements to public health and the environment. To all individuals and organizations who are working for a cleaner and more sustainable planet.

*"When written in Chinese, the word 'crisis' is composed of two characters. One represents danger and the other represents opportunity."*

— John F. Kennedy

# CONTENTS

## Part Three: Imagining Tomorrow

# LIST OF FIGURES

# DRIVING THE FUTURE

# INTRODUCTION

*8:00 p.m., Beijing, China*

On September 30, 2040, a slight woman in jeans and a T-shirt swipes off her e-book and steps out of a silently idling car on a busy street.

"Park Mode," she says and then watches as the empty, single-occupant car pulls down to the entrance of the parking garage underneath her apartment building. The woman doesn't have a license—she normally takes the train and shared taxis around the city—but decided to pay for a car on demand to drive her out of town to a friend's house for the weekend. After pausing for an exiting electric scooter, the driverless vehicle disappears into the lot.

The car's surface still glows in the woman's favorite shade of purple. The color-changing exterior also acts as a solar panel that powers the air conditioning and accessories when the vehicle is idling in traffic. The car's body contains no steel, but is constructed primarily out of aluminum and plastics, weighing in at about 1,650 pounds. The electric engine gets about 100 miles per gallon (mpg), close to the global average, and puts out zero emissions. In 2030, the Chinese capital passed laws that allow only zero emissions vehicles within city limits.

The car's navigation system has helped it travel 120,000 miles without so much as a scratch, but the technology also helps the car quickly find a parking space in the darkened garage. Four low-light cameras integrated into the frame guide it into a cramped space between two other vehicles. The car's exterior fades back to its default blue shade, and it begins wirelessly powering up for the next customer.

### 1:00 p.m., Berlin, Germany

A seasoned German truck driver stops for a lunch break on the outskirts of town. He is almost done with his day's work, transporting motorcycle parts from Frankfurt in his two-trailer, autonomous truck. The truck is the latest generation truck from Daimler and sports a cellulosic, biodiesel hybrid system and is capable of driverless operation. It has been averaging a relatively constant 85 miles per hour (mph), making the 344-mile trip just over four hours, always connected to the "Connected Cloud Drive" system. The driver had been listening a lot of Karl May audiobooks on his trips recently and briefly fell asleep somewhere near Leipzig. But on the dedicated highway for autonomous trucks, his short nap wasn't any more dangerous than a passenger falling asleep on a plane. About fifteen minutes from the shared traffic routes, he had to punch in his driver code and grab the wheel. But he only had to intervene once when heavy rain set off a warning that the visibility of the truck's control cameras might be impaired. Even in that extreme weather, the truck was in continual communication with the vehicles around it as well as receiving information from the road's own traffic warning nodes to keep it on track.

### 12:00 p.m., London, United Kingdom

A young London City trader walks out of his office on the way to meet his girlfriend for lunch. He pauses to pick out a flower at a stand before pushing the valet app on his smartwatch. He typically does not drive in the city—central London has had a daytime congestion fee for vehicles since before he was born—but he can't resist a quick spin in his company's new executive share car, a shiny black 2040 model fuel cell 911 Targa. As it turns the corner, the car's almost entirely car-bon fiber body gleams, but the super-strong and lightweight material encases a vehicle that weighs less than the tiny 1954 Austin Healey coupe his dad used to drive on the weekends.

The Targa glides up, and the man slides underneath a roof made of new generation solar panels with a nanotechnology coating that produces even more electricity from the wind friction generated at higher speeds, refilling the battery packs as he drives. The car can travel a more than 400-mile range on a five-minute quick-fill charge of the solid-state battery pack and hydrogen tank. The fuel cell engine is virtually silent, but with the optional flywheel technology the car can accelerate from zero to 60 mph in around 3.5 seconds.

He forwards his girlfriend's text with the restaurant location to the car's navigation system. The Porsche transmits its presence on the road to the City, which transfers £30 from his company's account to the public transportation network. Then the vehicle pulls out behind a bus on the otherwise quiet street.

### 7:00 a.m., Atlanta, United States

The Atlanta suburbs are already a humid 90 degrees. As a middle-aged man stands in the shower, his car charges wirelessly in the garage while the air conditioner precools the interior. The vehicle is small, about the size of the Honda Fit he and his wife used to own, and can travel about 100 miles on a full charge of the advanced battery pack, which is the size of a large briefcase. At 7:15, he hops into his car and sees a few crumbs on the passenger seat. The whole windshield is capable of turning into a high-resolution screen; his daughters had snuck out there last night to play video games.

Just for the fun of it, he backs the car out of the driveway himself before allowing the steering wheel to retract. While the car takes him the five minutes to the train station, he has a quick video conference with the sales staff at his low-carbon fuel company. The car drops him off and drives itself home.

At 7:30 the two kids jump into their car, which takes them to school. Their mom is telecommuting today but has an appointment with her personal trainer at 12:30. The gym has a drop-off zone but no parking lot, so, after the woman hops out, the car waits in a designated space two blocks away.

Then, at about 1:00 p.m., the eighth grader calls in. She feels sick and wants to go home. The car texts the mom, asking for permission to pick her daughter up, and then texts the girl. The car knows that the fitness appointment will run another 30 minutes, so it picks up the daughter before returning to the gym. Later in the day, the car also picks up the other child from school, along with her friend who is coming over to study. After charging for fifteen minutes, the car picks up the dad from the train station. Even though the family has three people of driving age, the autonomous car allows them to get around with just a small urban commuter car. That weekend the whole family is taking a trip to the beach, so they'll use some of their lease's contractual "shared vehicle days" for a larger plug-in vehicle.

### 6:00 a.m., Mexico City, Mexico

A couple that has just moved into a new apartment complex is fast asleep. In the past, their car purchases had been dictated by where they lived. They didn't need one in the crowded center city but decided they wanted a vehicle to get around once they moved to the outskirts. They still use public transport to get to work—the car is mostly for personal use. In this case, however, they chose their apartment because of their car. They save money on their home's power bill by plugging their electric vehicle into their apartment's "green circuit." Right now, during lower use hours, the battery pack is storing the cheaper wind-generated electricity that the local microgrid supplies. During the day, the microgrid switches to solar. Later, when it is sitting unused in the garage during the daytime peak load hours, the battery pack provides lower cost, clean electricity for their apartment.

★   ★   ★

How close will these imaginary scenarios of the future be to reality? For half a century, the promise of futuristic technologies—high-tech computer cars, electric and driverless cars—has been dangled in front of us in magazines and television shows. But a yearly trip to the dealership would show only modest technological improvements in new models.

Today, those promises are finally being realized. In 1970, for example, there were zero mass-produced hybrid electric vehicles. The same number of hybrids was available to drivers in 1980 and the early 1990s. By 2000, there were two hybrid models. Today there are over fifty.

Even more amazing are the growing number of pure electric cars and hydrogen fuel-cell vehicles that use no gas at all. Several of these vehicles already get as much as 100 mpg today, a number that, until recently, would have seemed ripped from the pages of science fiction. Increasingly smart and interactive vehicles with functions like lane change assist are already available in multiple production models. Several prototypes for driverless cars are being road tested. The automobile market has changed dramatically and irrevocably. The future is now.

This transformation is far from over, though. Over the next decades, the proliferation of alternative fuels and powertrains will make cars many times quieter, cleaner, more efficient, and smarter than they have ever been. In the near future, we will see automobiles that not only are high efficiency or emissions free but also completely reimagine personal transportation. More and more people will get around via automobiles that have more in common with cell phones or airplanes that our traditional concept of a steel box with a gas-burning engine. In short, we are in the early stages of a transportation revolution bigger than any since Henry Ford began mass production of his Model-T. But just as remarkable is why this sea change in automobile design is taking place now.

The conventional automobile is one of the twentieth century's great success stories. But this same petroleum-burning technology that radically increased personal mobility came with a huge cost. By the 1940s and 1950s the automotive exhaust in American cities was sickening citizens and darkening skies. By the 1960s, the smog in Los Angeles had become a poster child of rampant pollution for the nascent environmental movement. By the early 1970s, this environmental consciousness pushed the federal government to begin cleaning up conventional pollutants in automobile emissions with new technologies and improved fuels. By 2010, new technology and fuels had resulted in cars that emitted 99 percent fewer conventional pollutants. But today's revolution in automobiles is designed to face down our biggest environmental challenge yet, climate change.

In addition to foul-smelling exhaust, cars—along with trucks, boats, planes, and trains—have been releasing a threat that is undetectable to our senses. The powertrains of the transportation sector run almost exclusively on petroleum. And every gallon of gas or diesel that they burn emits carbon dioxide, the greenhouse gas that accelerates climate change. As a result, about a third of the total greenhouse gas emissions in the United States are from the transportation sector. Globally, the sector accounts for about a quarter of greenhouse gases, but transportation is also the fastest growing contributor to overall greenhouse gas emissions.

This growth has many factors, but among the most important is the purchasing power of an Asian middle-class swelling from about 500 million in 2012 to around 1.75 billion people by 2020, a 3.5 times increase in only eight years. In 2013 the Chinese consumer alone bought more than 22 million vehicles, and in 2020 they are expected to buy more than 30 million vehicles. That is 10 percent more than the 2013 combined sales in North America and Western Europe. To avoid a climatic catastrophe, the enormous number of new cars on Chinese roads alone must be significantly cleaner. The good news is that China, like other major economies, is moving aggressively to reduce transportation emissions. The global efforts to rein in greenhouse gas emissions in the transportation sector have been much more successful than most other international treaties targeting the existential threat of climate change.

★ ★ ★

Over the past three decades, the dangers related to climate change have become increasingly clear and immediate. Ocean levels and surface temperatures have been rising, and each of the last three decades has been successively warmer at the earth's surface than any preceding decade since 1850. In the form of devastating droughts, extreme flooding, heat waves, and fires, the earth got a taste of the destructive changes that would become increasingly common as climate change accelerated in the near future. While climate scientists don't like to connect any one event to global warming, their modeling conclusively shows that humans can expect more of the same weather as an unprecedented growth of greenhouse gases in the atmosphere continues to drive global warming.

Unfortunately, the world's demand for action on climate has been met with ineffective treaties and half-measures by the global community. Worse, the United States refused to support international treaties and has been unable to pass any federal domestic legislation targeting climate change. For decades, the oil, automobile, coal, and power generation industries took a page from the tactics used by the tobacco industry in the 1980s to cast doubt on the climate science and to scare the public about the economic costs of action. The cost of this intransigence is estimated by the Center for American Progress at US $188 billion in climate-related damage in 2011 and 2012,[1] and it increases by the day. A 2014 White House report suggests that every decade of inaction actually increases the cost of mitigation by 40 percent.

Then, in the first year of the Obama administration, an opening presented itself, and the United States enacted the first greenhouse gas reduction regulations in its history. The 2012–2016 and the 2017–2025 Light Duty Vehicle greenhouse gas and CAFE rules will halve the nation's carbon emissions from cars and double their average fuel economy to 54.5 mpg by 2025. When the *Economist* examined twenty global greenhouse gas reduction actions to combat global warming in the article "The Deepest Cuts" (September 20, 2014), comparing their respective contributions to greenhouse reductions, the United States vehicle rules ranked as the most significant action in the transportation sector to date, worldwide. These same historic regulations are also driving the design of a new generation of cleaner, smarter cars.

A number of factors aligned to create this groundbreaking regulation including a Supreme Court decision, a California state law, a suddenly bankrupted auto industry, and an administration determined to do something about climate change. It was this perfect alignment of events that gave my office at the EPA a real chance to finally rein in the greenhouse gas emissions that gas-powered cars have been emitting into the atmosphere ever since they first hit the road. Just as important, a converging set of emissions and greenhouse gas regulations being implemented across the world's major economies—including Europe, Japan, and China—promises similar reductions by 2025. These global regulations are among the key drivers required to combat climate change by changing the cars we drive.

★   ★   ★

The first part of this book tells the story of humans' growing understanding of science behind climate change and how, starting in the 1980s, a collection of corporate interests ferociously beat back national efforts to address the problem. It traces the history of the environmental policymaking in the United States along with automakers' foot-dragging and downright refusal to bring their most innovative new technologies to the marketplace.

Then we follow the efforts of a wide-ranging group of people—from a staunchly Republican Texas hedge fund millionaire to a former California public school teacher to the Georgetown lawyer who prepared the winning argument for a Supreme Court decision on greenhouse gases, to dedicated EPA engineers in Ann Arbor—as they play critical roles in finally breaking through this corporate and political stranglehold. It also describes how government, despite the barrage of negative reporting, can work to save and improve the lives of its citizens.

Next, we examine the primary drivers that are shaping, and will continue to shape, the car of the future while fighting climate change. Many of tomorrow's technologies, from alternative powertrains to lightweight materials to low carbon fuels and driverless car technologies are already in development. We'll explore their possibilities as well as innovative possibilities that sit on a more distant horizon.

In the final section I assess the scale of the challenge ahead for 2050, putting hard numbers on how much greenhouse gas emissions from cars and light trucks must be reduced to avoid the truly catastrophic consequences of climate change. I will lay out the additional  policies and innovations in automotive technology that will be required, while sharing some of the lessons I've taken away during my experience with environmental policy over the last three decades.

I consider this book a continuation of my public service, for it brings together the science, policy, and technologies that combat climate change with the cleaner, smarter, safer, and more exciting vehicles that our world needs. For the past forty years, the United States has led the world in innovative clean air technology. I remain optimistic that we will once again successfully confront the major environmental threat of our times—and do it while advancing Americans' health and our national economy. As a mother, I believe we have a moral responsibility to act urgently to protect our children's future.

# PART ONE

# THE CLIMATE JOURNEY

## I

# THE DAWN OF AUTOMOBILITY

In August of 1888, a thirty-nine-year-old woman named Bertha rolled to a stop in the center of Wieslen, a small town that sat in the rolling green countryside of southwestern Germany. She and her two teenage sons climbed out of their three-wheeled vehicle, made their way into the local apothecary, and purchased all the ligroin he had, about three liters. Ligroin is a petroleum ether commonly used for painting and decorating, but Bertha had neither of these uses in mind. Instead, she walked over to the carriage and poured the liquid into an engine sitting on the front. She then hoisted herself back into a wooden driver's seat and, to the surprised looks of pedestrians, chugged out of town with her family at the leisurely pace of six miles an hour.

The trip on the curious carriage was hardly trouble free. Along the way Bertha pulled over at a blacksmith's to fix a chain and a cobbler's to have him attach some makeshift leather brake pads. She'd had to solve other issues herself, using a hatpin to clean a clogged fuel pipe and a garter to insulate a wire. But, just before dusk, she pulled up outside of her mother's house in Pforzheim and, after unloading her sons, sent a telegram back to her husband in Mannheim. The message told him that she had stolen his prototype vehicle, had safely completed the sixty-six-mile journey, and would return home the next day.

The trip back was relatively uneventful, but the sight of the engine-powered vehicle still surprised and alarmed observers—exactly the reaction Bertha wanted. The trip's main purpose was to promote the automobile into which she and her husband, Karl Benz, had invested enormous amounts of money and time.

Or so the story goes. Over time, the legend of Bertha Benz's inaugural trip has been told and retold, embellished and celebrated by German drivers who retrace her journey every year. What is certain is that Karl Benz invented the first practical gas-powered vehicle, that Bertha's trip was arguably the most successful automobile promotional campaign to date, and that the Benz's internal combustion vehicle still faced stiff competition from other propulsion techniques.

At the turn of the twentieth century, popular automobiles were also powered by electricity and steam. Each technology had its own upsides as well as major shortcomings. Electric cars were quiet and easy to operate, but they were heavy, their distance was limited, and their recharge time was unbearably long. One of the most popular vehicles at that time was the Columbia Runabout, which could go forty miles on a single charge and run at speeds of up to 12 mph. Steam-powered cars accelerated fine once you got them going, but they took a long time to harness enough power to start. They also had limited capacity to store energy on long trips and were prone to explosions. Gas-powered vehicles were relatively quick starting and could run for long distances without refueling. But they were noisy, complicated to operate, and often broke down. By the early 1900s, there were nearly four thousand steam, electric, and gas-powered cars traversing America's roads. Which machinery would win out in the end was still anyone's guess.

Then, twenty-six years after Bertha's journey, Henry Ford's factory took automobile production to the next level. By 1914, his factories were churning out a boxy, black, gas-powered Model T every ninety-three minutes, completing six vehicles in roughly the time it had taken Bertha to go sixty-six miles. By 1918, the assembly-line manufacturing had also brought the car's cost down to $450, a price even within the reach of Ford's factory workers—and well below the $2,000 for electric vehicles. The Model T was so popular that Ford didn't buy any advertising for years. By the 1920s, electric and steam vehicles were left in the petroleum-reeking exhaust of the now-dominant internal combustion engine.

★   ★   ★

Of course, the popular availability of cars did much more than make Henry Ford rich. The automobile pushed horse-drawn carriages to the wayside and forced cities to pave roads—if not completely redesign them. It created the sprawling suburbs of 1950s America. It rapidly changed social mores and became one of the most desired consumer goods in the world. It also pushed a century-old revolution in personal mobility to a new zenith.

Over the past two centuries, while the earth's population has increased seven times and global GDP has increased a hundred times, personal mobility has increased a thousand times.[1] In 1800, people primarily got around by walking, riding horses, or using boats. The vast majority never ventured far from their homes or did so rarely. By 1950, millions of people could almost effortlessly travel hundreds of miles in a day. The internal combustion engine was the most important technology in this transformational increase in personal mobility.

Unfortunately, this agent of personal freedom was also a source of an enormous amount of air pollution. Beginning in the 1940s, people in urban areas throughout the United States experienced waves of then-unexplained pollution that caused teary eyes, headaches, nausea, asthma attacks, and other reactions. School children in Los Angeles were kept indoors during high-risk days. The filthy air that factories and power plants emitted was well known, but the smog that enveloped America's cities was not a sooty black; it had a brownish, yellowish tinge, which was due to the nitrogen oxide in auto emissions. The problem worsened until the popular disgust over pollution rattled American politics and eventually spilled over into one of the nation's largest grassroots demonstrations.

★   ★   ★

In 1959, a newly elected Democratic senator began his journey toward becoming the national political face of the new environmental movement. The former governor of Maine, Edmund Muskie, had grown up the working class son of Polish immigrants in the small town of Rumford. At six feet four inches, with a powerful voice, Muskie was soon known in the Senate for both his flashes of temper and his ability to find consensus.

When Muskie arrived in Washington, DC, what we call environmentalism today didn't exist. Hunters and fishermen like Muskie promoted conservationism by protecting their recreational areas. Forestry officials managed logging and mining interests in national parks. But there was very little concern over the effects pollution might be having on public health and the environment. Muskie had experienced the ill effects of pollution firsthand. His hometown was dominated by a paper mill that had been spitting enormous amounts of chlorine, nitrogen, phosphorus, and other pollutants into the water, while spewing sulfur dioxide and other toxins into the sky. Dead fish floated in the water, and the air pollution was so bad that it prevented new businesses from opening in Rumford. The new senator recognized the urgent necessity of protecting birds, frogs, humans, and other living things from pollution.

Throughout the 1960s, Muskie's modern-day environmentalism developed a broad base of support, from scraggly college students to middle-class moms worried about their children's health. Mainstream media's coverage of environmental disasters, like *Time* magazine's 1969 story of a fire on the surface of Cleveland's polluted Cuyahoga River, further galvanized concern. As a result, the first nationwide Earth Day events in 1970 attracted attention from across the political spectrum. Senators and congressman from both parties flew back to their districts to speak at local events in city parks and on university campuses. Meanwhile, Muskie was pushing through a bill designed to satisfy the new environmental concerns.

During the 1960s, a series of laws established increased federal monitoring of pollutants, but none had the authority necessary to begin the real work of cleaning up the country. By 1970, Muskie had created legislation designed to transform America's relationship with its increasingly polluted landscape. His bill also satisfied and incorporated various important political interests. From Tennessee Republican senator Howard Baker he included a commitment to technology-forcing regulations—rules that don't prescribe specific solutions for industry to follow but stimulate the marketplace by creating a demand for new technologies. Muskie also respected Missouri Democratic senator Tom Eagleton's demand for solid deadlines. Finally, Muskie wanted certain health standards to be met. These elements all fused in the creation of the most effective environmental law to date, the Clean Air Act of 1970. The bill locked in the regulatory technique of using technologies to achieve a healthier environment by a certain date. No other federal law combined all these elements—most had none.

A well-known example of the Clean Air Act's application quickly followed its passage. By 1975, new cars were required to meet certain standards for emissions. The regulation would dramatically clean up the air, reducing cases of bronchitis, asthma and more serious health effects including premature death. The requirements had to be met by 1975 model year cars. And it was up to the car companies to figure out what technology would get the job done, prompting a race between multiple companies to supply a device that would win the millions of dollars in contracts from auto companies. That solution, the catalytic converter, is now found in virtually every new car manufactured in the world.

But Muskie wasn't the only one responding to the new environmentalism. Soon after the Clean Air Act became law, President Richard Nixon created the Environmental Protection Agency. A

Republican and no devout environmentalist, Nixon had essentially sat out Earth Day in the 1970s. But he did respect the political importance of the widespread public demand for cleaner land, water, and air. In fact, the EPA may have been strategically created in part to take the environmental card away from Muskie, Nixon's presumed Democratic challenger in the 1972 presidential election. A few months after the creation of the EPA, the Nixon campaign forged and leaked a letter to the press that called Muskie's credibility into doubt, essentially ending the environmentalist senator's presidential ambitions. But whatever Nixon's motivation was, the new agency he created was too powerful to be considered one more political trick. The EPA soon exercised its authority under the Clean Air Act to begin a massive cleanup of the mess that internal combustion–powered automobiles had created over the past half-century.

The EPA's success in this effort is clear when comparing modern cars with those from 1970. Just forty years after the EPA went to work, cars produce over 99 percent fewer emissions than they did in 1970. Between 1970 and 2010, the reductions in levels of soot, smog, and other pollutants from cars, trucks, factories, and power plants have prevented hundreds of thousands of premature deaths, nearly a million cases of chronic bronchitis, and over 18 million child respiratory diseases. Lead reductions have prevented the loss of over 10 million IQ points in children.[2] Americans breathe much cleaner, healthier air than they did in 1970. While our cities' skies may not be pristine, they are remarkably cleaner.

★   ★   ★

When I joined the EPA in 1980, it was my first professional experience with environmentalism. I had arrived in the United States from Athens, Greece, twelve years earlier, as the environmental movement was becoming a powerful political force. When I got to America my English was limited, but, after a summer-long crash course, I started an engineering program at the University of Massachusetts in Lowell.

At that point my knowledge of environmental issues was also pretty limited. Growing up in Athens, a crowded and polluted city that is home to a third of the entire country's population, I was constantly surrounded by atmospheric pollution from carbon monoxide, sulfur oxides, nitrous oxides, and dust. I watched Athens' air pollution coat the 2,500-year-old marble pillars and statues of the Parthenon in soot. I had seen and read about how the beautiful marble underneath the soot was being eroded by pollution, which made the future of the monument questionable.

After receiving my MS degree, I went on to a job at a Connecticut-based consulting firm for the chemical industry. It wasn't until I started at the EPA's Office of Toxic Substances that my work as a chemical engineer became exciting. I joined the EPA team evaluating the toxicity level of new chemicals before they entered the marketplace. Now my engineering skills weren't just academic; they were preventing toxins from harming people and the environment.

In 1985, I secured a one-year detail to work for Senator John Chaffee, a Republican representing Rhode Island. Chaffee, the powerful chairman of the Senate Environment and Public Works Committee, helped me to author a bill that made plastic rings for six-pack beverages biodegradable. At the time, every six-pack of soda or beer was held together by plastic-ring packaging. A lot of these rings found their way into the ocean, where they trapped and killed marine life. The bill I worked on became part of a Superfund amendment in 1986. It was my first experience with the process of lawmaking.

From the Senate, I returned to the Office of Toxic Substances as a deputy director. As a manager, my first project was the toxic release inventory initiative, which became better known as the "Community Right-to-Know Program." For the first time, the program ensured that public could learn about the type and amount of toxic substance releases from facilities in their community. Several years later, I took on the job of directing the Office of Radiation and Indoor Air. Among my responsibilities was helping to establish the first repository for nuclear waste in Carlsbad, New Mexico. I also worked to make radon gas and other indoor pollutants' health impacts known to the public. With my team, I also managed the publication of the first scientific study about the health impacts of secondhand smoke, which completely changed the policies for smoking in public places.

Then, after moving from various positions over my first fourteen years at EPA, I found a home leading the Office of Transportation and Air Quality. It was here that I directly confronted the legacy of innovators like Henry Ford and Karl Benz. Starting in 1994, I spent the next eighteen years working with my team to reduce the air pollutants that caused premature death and respiratory illness from cars, commercial trucks, and buses. We reduced cancer-causing substances like benzene in gasoline and sulfur in diesel and gasoline fuels. We also developed rules compelling locomotives and boats to reduce their soot- and smog-forming pollutants that impacted the health of millions of Americans.

But while my team successfully pushed forward our agenda to clean up and reduce conventional air pollutants up to 99 percent since the seventies, a mostly odorless, invisible threat was rapidly building up all around us, greenhouse gases. Throughout the 1980s and '90s, as Americans became more familiar with this threat, several things became clear. First, human-produced greenhouse gases were causing the planet to heat up rapidly. Second, along with other greenhouse gases, one of the major culprits in this process was carbon dioxide, a byproduct of burning fossil fuels like diesel, gasoline, and coal. Third, greenhouse gases were streaming out of the tailpipe of just about every car and truck in America. In fact, the transportation sector—the sector my office was responsible for cleaning up—was producing almost a third of the greenhouse gas emissions in the United States, second only to the power generation sector.

Despite all of our efforts at EPA, an unregulated, life-threatening gas was still streaming out of vehicles across the country. Over its first forty years, the EPA accomplished much of its original purpose.

But it turned out that our greatest challenge still lay ahead: the global threat of greenhouse gas–driven climate change. This phenomenon was a complex beast with a global reach—and a problem so politicized that it had been dangerously ignored by our government for decades.

# THE DISCOVERY OF EARTH'S CLIMATE

Kiribati, a line of South Pacific islands strung out in turquoise water, has long been a poor nation, subsisting on international aid and the sale of its fishing rights to wealthier countries. But over the past decade, the country has begun to face a more existential threat. Rising sea levels are causing the tropical waters to flood churches and gobble up homes near the coast. Further inland, briny liquid pours out of the ground, collecting in fetid pools. Stands of palm trees are dying from exposure to salt water. The citizens of Kiribati navigate these new challenges on a day-to-day basis, haunted by the realization that their whole world is collapsing. The gorgeous archipelago has become a stark image of climate change's nightmare scenario.

The nearly complete disappearance of the nation by the end of the century has become so certain that the president, Anote Tong, spent nearly $10 million to purchase 6,000 acres in Fiji. His people don't want to move, but they will need somewhere to live and, within a generation or so, thirty-two of the thirty-three islands will be uninhabitable.[1] The message to the rest of the world is clear: global warming is no longer limited to a series of dire predictions about what might happen decades in the future. It is happening now.

For Americans, it may be easy to overlook the fate of a hundred thousand people trapped on a series of sinking ships in the South

Pacific. But what about South Beach? Today, this fashionable, wealthy strip of Miami is under siege. Rising waters cause slimy green liquid to bubble up from gutters even when it is sunny out. Shopkeepers have plastic bags and rubber bands on hand for people to cover their shoes. First floor garages flood with seawater, corroding the cars inside. The city is spending $1.5 billion to try and hold back the water, but the odds of any lasting success are slim.

The mayor of South Miami, Philip Stoddard, says that just a one-foot rise in sea level wouldn't actually wash away houses but would still destroy the city's infrastructure. Residents would no longer be able to flush their toilets or get fresh water from the sinks. At that point, anyone who can afford to will move out and property values will crash. "It makes one thing clear, though," the mayor says. "Mayhem is coming."[2]

A University of Miami geology professor named Harold Wanless agrees that, in a place where 2.4 million people live at a maximum of four feet above sea level, it's no longer a question of if but when. "It is over for south Florida. It is as simple as that."[3]

Miami is far from alone. Seven hundred miles across the Gulf of Mexico is New Orleans, half of which sits under sea level. After the storm surge from Hurricane Katrina breached the scores of levees that surround the city in 2005, flood waters covered almost the entire city, killed upwards of eighteen hundred residents and so crippled the infrastructure that corpses lay in the streets for days. Up the Eastern Seaboard, New York City has flooded repeatedly from hurricane storm surges. Most recently, 2012's deadly Hurricane Sandy closed the subway, school system, and the New York Stock Exchange for days. As a result, the city has already developed a comprehensive plan to combat future flooding, including a series of up to twenty-foot levees.[4] In fact, the whole United States itself is literally smaller than it was fifty years ago, as beaches and barrier islands are slowly gobbled up by rising water.

In 2003, Europe faced another type of extreme weather, a catastrophe that, strangely enough, remains relatively unknown in the United States. That summer was the hottest recorded in centuries across the continent. The United Kingdom, Scotland, and Switzerland recorded their nation's highest ever temperatures, causing thousands of deaths as well as avalanches and flash floods that tore through the Alps. Fires ravaged Portugal, burning up 10 percent of the country's forests, while ships in drought-stricken Germany could not navigate the Elbe or Danube Rivers. All told, roughly 70,000 people died as a result of heat stroke, dehydration, organ failure, and other causes related to high temperatures. Along with flooding, heat and extreme drought are the weather phenomena most closely associated with climate change. As climate change accelerates, these types of extreme weather events will become more common.

In fact, every year there are new indications we are moving through the early stages of a climatic crisis. The prolonged droughts, extreme heat, and severe floods experienced by the country and the rest of the world over the past few decades are all hallmarks of global warming-induced catastrophes. Of course, extreme weather events have always come and gone, but longer-range indicators also point in the same direction. The eight-inch rise in global sea levels over the past century is a broader gauge of where we're headed. The fact that each of the previous three decades has successively been warmer than any previous decade since 1850 is another.[5]

The increased acceptance that global warming is severely impacting the world has also created new heralds of the oncoming crisis. The US military, for example, looks at climate change-induced crises as "threat multipliers" that are likely to bring down governments, destabilize entire regions and encourage terrorism. After being hammered by hurricanes and other huge climate change-related storms, insurance companies have dramatically raised rates on coastal property. Even Exxon, the long-time funder of climate change deniers, has issued a generic acceptance that climate change poses risks. More important, the company's own scientists are studying the commercial ramifications of the weather phenomenon on their business, such as increased opportunities for deep sea drilling operations as larger swaths of the Arctic Ocean become ice-free.

In June 2014, a high-profile trio drawn from the upper reaches of politics and finance, former US treasury secretary Henry Paulson, former New York mayor Michael Bloomberg and billionaire investor Tom Steyer, released *Risky Business*, a report laying out the potential economic losses to the United States from global warming.[6] *Risky Business* was not another of the increasingly dire scientific announcements about climate change. In fact, there was no new science at all. Instead, page after page laid out the billions of dollars the United States economy stood to lose unless businesses reduced their greenhouse gas emissions. By 2050, the report suggested, up to $106 billion of the nation's coastal property will likely be below sea level. By the beginning of the next century, that loss could be half a trillion dollars.

Though the issue still marks a fierce partisan divide, both the Bush and Obama administrations have issued reports evaluating the specific national risks of climate change's impact on the nation's economy, health, and national security. The most recent, the 2014 National Climate Assessment, provides a clearly written overview of how America's dynamic climate has and will drive radical changes to all aspects of its citizens' lives.

In the future, the report states, the ground in some regions will be parched from drought more often. In coastal Alaska it is washing away. Flooding rivers will overwhelm water control systems more often. The oceans are not only warmer but too acidic for oyster farmers from the Pacific coast to New England to raise healthy stock. The winters will be shorter. The air is filled with more allergens. The summers will be unusually hot for extended periods of time lasting longer than any living American has ever experienced. Since 1895, temperatures have risen around .85 degrees Celsius (1.5 degrees Fahrenheit), with most of the change happening since 1970. Over the next few decades, that temperature change is likely to increase another two to four degrees Celsius (three and a half to seven degrees Fahrenheit).[7]

★   ★   ★

Just as climate scientists' predictions of global warming are now a living reality, the cause of these earth-changing deviations should also no longer be subject to debate. Over the past two centuries, humans have pumped huge amounts of additional greenhouse gases into the atmosphere, causing the earth's atmosphere to retain more of the sun's heat and rapidly warm the planet in the process. There are a number of heat-trapping greenhouse gases, including methane and nitrous oxide, but the single biggest contributor to global warming is carbon dioxide.

Electricity production, which most frequently involves burning coal, oil, and other fossil fuels, is the single largest contributor of additional carbon dioxide. Transportation, the industry that I regulated for nearly two decades, ranks second. But transportation is also the fastest growing carbon-producing sector. While we certainly need to rein in the carbon emissions of electricity, dramatically reducing the carbon dioxide emitted by cars, trucks, trains, planes, and ocean-going vessels will be enormously important in putting the brakes on global warming.

Despite the lack of action that has marked the United States' response to global warming, the facts are simple: the world is heating up because of human activity. Without humans on the planet, the earth would currently be in a slight cooling trend instead of a period of significant warming.[8] Climate scientists are constantly updating their precise forecasts of the future, predicting how many more islands and coastlines the sea will invade or what the total cost of agricultural damage caused by severe droughts will be. But, though it has been debated and obfuscated by climate change deniers for decades, the big idea that we are radically and dangerously changing our planet's weather is not new. In fact, the modern understanding of global warming only emerged after a century and a half of globe-spanning scientific investigation.

★　★　★

Questions about how human activity might affect the weather actually go back millennia. Several thousand years ago, my Greek ancestors debated issues such as how cutting down forests might change rainfall in a certain region. Of course, the ancients Greeks didn't have enough data to prove anything definitive, even locally. The real beginning of what became modern climate change theory dates back only a few hundred years when some nineteenth-century visionaries began the hard task of convincing the scientific establishment that the earth's climate went through dramatic changes at all.

At the time, the generally accepted assumption was that the earth had been locked in a process of continual cooling ever since its fiery birth billions of years previously. Ice ages were unknown. The periodic heating and cooling of the earth was not even a serious theory. So, before considering the possibility that mere human action could change the earth's temperature, someone had to prove that anything did.

In 1835, while studying the moss growing on boulders in a mountainous region of Bavaria, the thirty-two-year-old German poet and botanist Karl Friedrich Schimper became perplexed as to how these enormous rocks had ended up sitting in the green foothills of the Alps.[9] They seemed erratically placed, weren't likely to have rolled from somewhere, and certainly couldn't have been moved there by people. These slabs, weighing many tons, rested in the mountains as massive mysteries of nature.

Schimper wasn't the first to run into this puzzle. For decades, close observers of nature—including the famous German author and botanist Johann Wolfgang von Goethe—had asked the same question. Neither was Schimper the first to conclude that the huge rocks must have been deposited on the vibrant landscape by huge sheets of ice. Glaciers were certainly strong enough to move the slabs. But since there was no such ice nearby, Schimper decided the frozen mass must have retreated up the hills ages ago. This assumption presented another problem. Such a theory could only be explained by an icier, much colder period in the earth's past. Schimper's novel accomplishment was synthesizing naturalists' previous work with his own fieldwork, formalizing solid theories, and then gaining support among other scientists.

Over the next year, he gave several lectures in Munich suggesting that, rather than undergoing a constant cooling off, the earth's climate was more dynamic. In 1837, working with other scientists, including the Swiss geologist Louis Agassiz, Schimper developed a theory of sequential glaciations and coined the phrase "*Eiszeit*"—"ice age" in English. But when Agassiz presented this groundbreaking work at a conference of the scientific establishment, he met with staunch resistance. Few scientists were willing to break from the accepted "volcanist" theory of continual climatic cooling. Ironically, this conference was held at a university on the shores of Lake Neuchâtel, a body of water that had been created tens of thousands of years earlier by the very process of glaciation that the bulk of scientists present denied.

Despite this setback, proponents of the ice age theory embarked on geologic fieldwork to bolster their arguments. After another four decades of establishment resistance, James Croll's *Climate and Time, in Their Geological Relations*, published in 1875, marked the general acceptance of ice ages as a feature of earth's dynamic climate. Of course, now that most scientists believed that the earth occasionally experienced much colder eras, they needed to explain how this temperature phenomenon reversed itself. After going into a deep freeze, what was the earth's mechanism for heating itself back up? The explanation for this phenomenon required another nineteenth-century scientific breakthrough, this one in chemistry.

★   ★   ★

By the mid-1800s, scientists had calculated that the sun's radiated energy was insufficient to heat the earth to its average surface temperature. To explain this discrepancy, several advanced the idea that the planet didn't just act like a huge piece of rock sitting about 92 million miles away from its heat source. Instead, the earth's atmosphere must trap some of the sun's energy. They were headed in the right direction, but it wasn't until the Irish scientist John Tyndall was able to separate individual atmospheric gases that the theory had any empirical proof.

In the late 1850s, Tyndall created a spectrometer that allowed him to run various gases through a series of heated tubes and then record their temperatures. As it turned out, the odorless, colorless gases Tyndall tested reacted very differently with the radiated heat. Water vapor and carbon dioxide, for example, retained much more heat than oxygen, nitrogen, and hydrogen. Tyndall's experiments confirmed that the atmosphere sponged up some of the sun's heat. They also highlighted those gases that were much more important than others in keeping that radiative heat near earth. Without water vapor in the earth's atmosphere, Tyndall wrote, the earth would be "held fast in the iron grip of frost."[10]

By the end of the 1800s, scientists had determined that the earth underwent enormous swings in temperatures and that gases like carbon dioxide and water vapor kept the planet warmer than it otherwise would be. In the first decade of the twentieth century, a scientist would put these pieces together to create an early version of global warming theory.

★   ★   ★

Svante Arrhenius, a Swedish physicist with a walrus mustache and the 1903 Nobel Prize in Chemistry, built on Tyndall's work by assigning mathematical values to the warming effect of these heat-retaining gases. But instead of water vapor, Arrhenius focused on carbon dioxide. According to his calculations, an atmosphere without that gas would result in earth's surface temperature falling about 21 degrees Celsius (38 degrees Fahrenheit). He also recognized the knock-on effect: a cooler atmosphere couldn't hold as much water vapor, resulting in an additional temperature decrease of approximately 10 degrees Celsius (18 degrees Fahrenheit). In other words, without atmospheric carbon dioxide, the earth's average temperature would fall below freezing.

Taking the opposite tack, Arrhenius also produced a formula calculating how increases in atmospheric carbon dioxide would change surface temperatures on earth. He determined that "temperatures of the Arctic region would rise about 8 to 9 degrees Celsius [14 to 16 degrees Fahrenheit] if carbonic acid increases 2.5 to 3 times."[11] With this data in hand, Arrhenius made his most critical contribution to modern climate change theory. Human activity, he claimed, might already be affecting this equation, increasing the earth's temperature.

Arrhenius's theory was, no doubt, encouraged by the pervasive evidence of the radical environmental changes wrought by the Industrial Revolution. For Arrhenius and any other European, the visible impact was everywhere. Rural areas were marked by clear-cut forests, while unfiltered black fumes from factories drifted through cities, staining buildings and coating lungs with heavy layers of soot. In 1873, a week-long inversion over London trapped industrial smoke near the ground, killing seven hundred citizens and bolstering the city's new nickname: "The Big Smoke."

Many Europeans thought that all this filthy pollution caused by the voracious consumption of coal, wood, and oil was the nasty price of progress. Arrhenius also imagined the invisible impact that the unprecedented amounts of dirty fossil fuels were having on the atmosphere. Every time a business or home burned a piece of coal and log of wood, the fire didn't just give off smoke; it released carbon dioxide into the air. With these processes now happening on a massive scale, Arrhenius believed that human-produced carbon dioxide might be able tip the level of atmospheric carbon dioxide enough to cause the planet to heat up. He was ecstatic.

In his 1908 book, *Worlds in the Making*, Arrhenius popularized the idea that a "hot-house theory" would allow humans to "enjoy ages with more equable and better climates, especially as regards the colder regions of the earth, ages when the earth will bring forth much more abundant crops than at present, for the benefit of rapidly propagating mankind." Arrhenius even thought that carbon dioxide emissions might help push back a future ice age.[12]

Though he was no environmentalist in the modern sense, Arrhenius's search for an explanation of temperature changes in the upper atmosphere resulted in the most complete, nontechnical introduction to modern climate change theory to that date. Despite his positive interpretation of the phenomenon, Arrhenius had opened the door to predictions about the invisible negative impacts of carbon dioxide buildup. In a time before automobile ownership was widespread, he provided the tools to predict the future impact of petroleum-burning emissions.

However, the ideas Arrhenius advanced about the earth as a hot-house didn't advance beyond the theory stage, in part because scientists underestimated the rapid buildup of carbon dioxide. Just as with the original ice age theory, most scientists preferred to argue on the side of relative stasis in earth's climate. Humans, they said, couldn't possibly create enough carbon dioxide to change the global atmosphere. The vast oceans, for one, would suck up far more carbon dioxide than all of human activity could produce. As a result, the first widely known theory about human-driven global warming sat largely discredited and forgotten for decades. It wasn't until the end of World War II that new research techniques and a scientific spending spree by the American military helped rehabilitate Arrhenius's work.

★   ★   ★

In 1946, a small army of scientists dispatched by the US Navy invaded the Bikini Atoll in the Marshall Islands in advance of planned nuclear testing. The Navy wanted scientists to investigate the potential for tsunamis or the destruction of fisheries during the bomb detonation tests. But during the official research, one of the expedition's leaders was also able to send off some scientists to do studies on a personal interest of his, the chemistry of the atoll's waters.[13]

Roger Revelle, an enthusiastic geologist and oceanographer, had risen to the rank of US Navy Commander during World War II, in part because of his adeptness at detecting enemy submarines. In the years following the war, Revelle had managed to maintain his close relationship with the Office for Naval Research, one of the military's most generous funders of scientific research. Through this lucrative connection, Revelle was able to build up the reputation and capacities of the Scripps Institution of Oceanography, a building that sat dramatically perched on the California coastline overlooking the Pacific. The Navy's funding also allowed him to pursue other interests. Bit by bit, the mysteries of water chemistry began to engage his time.

Finally, in 1957, he coauthored a paper laying out his research. Contrary to the arguments that had helped sink Arrhenius's work, Revelle suggested that the world's oceans were much less capable of absorbing the excess carbon dioxide in the atmosphere than was previously believed.[14] This was a major development. The oceans are the world's largest absorbers of carbon dioxide, what are known as "carbon sinks." But, if Revelle was correct, the water covering more than 70 percent of the earth's surface couldn't be relied on to prevent the climate from heating up as more and more fossil fuels were burned. Although it built on and helped validate Arrhenius's theory, Revelle's discovery dissented on interpretation. Far from sharing the Swede's favorable impression of a buildup in atmospheric carbon dioxide, Revelle thought it was a serious climate threat.

Revelle was not the only scientist pursuing work that proved critical to the modern understanding of global warming. His 1946 excursion to Bikini Atoll came a year after World War II had essentially been ended by nuclear physicists. As the Cold War intensified, the United States military continued to pour money into many different scientific fields. The scramble to gain tactical advantages over the Soviet Union also provided rare opportunities for climate researchers.

For example, a post-Hiroshima world gave scientists a rare opportunity to track radioactive isotopes in the atmosphere, creating the first accurate mapping of the earth's wind currents. This work also showed researchers how locally produced carbon dioxide emissions were rapidly dispersed throughout the entire atmosphere. In other words, no matter where the carbon dioxide came from, climate change would only be a global event. Studies examining how infrared rays passed through the atmosphere were designed to improve heat-seeking missiles, but they also helped further our understanding of the mechanisms of radiation absorption by the air. Research teams were sent to the Antarctic because the military wanted to learn how to fight a war there, but the remarkably clean air was perfect for monitoring carbon dioxide levels.

All this new data lent new authority to Arrhenius's half-century-old theory. But while climate change began to look more real, it also continued to appear a lot more menacing than Arrhenius's sanguine predictions about increased crop yields and more pleasant temperatures. Now science needed a messenger to tell the rest of the world about the growing consensus around this new threat. Once again, the tireless Roger Revelle stepped in.

★  ★  ★

In 1957, Revelle authored the bureaucratically-titled *First General Report on Climatology to the Chief of the Weather Bureau*, the first of his efforts to urgently warn the government that humans' unprecedented consumption of fossil fuels amounted to a "grandiose scientific experiment."[15] He also testified before a Congressional committee, saying that burning fossil fuels might lead to dramatic and erratic shifts in the earth's climate within the next century. That same year, Sputnik was launched, kicking off fear that Soviets were pulling ahead of the United States—and increasing scientific funding opportunities.

Nine years later, in 1965, Revelle was part of the environmental team for the President's Science Advisory Committee that reported climate dangers to President Lyndon Johnson. At around the same time, his lectures at Harvard University inspired future senator and vice president Al Gore to adopt climate change as a signature issue. The science of the greenhouse effect and carbon-driven climate change was becoming increasingly sophisticated. More important, these ideas were beginning to circulate outside of exclusively academic circles. Still, there was no sign of action on the issue, aside from reports recommending more studies.

Then, in 1977, the *New York Times*'s crusading science writer Walter Sullivan brought the mechanisms of the greenhouse effect to public prominence for the first time. In a 1977 front-page story titled "Scientists Fear Heavy Use of Coal May Bring Adverse Shift in Climate," Sullivan discussed the findings of a blue-ribbon panel of scientists who warned that, among other dangers, carbon dioxide–driven climate changes could "radically disrupt food production and lead to a twenty-foot rise in sea level."[16] The article mostly revolved around coal and not other sources of carbon dioxide, although other sources would gain traction in the following decade. Within a few years, Sullivan and a widening group of journalists began regularly reporting on the increasingly conclusive data showing climate change was real and a serious threat. In 1981, Sullivan described new proof of an overall warming trend since the nineteenth century, setting the stage for mainstream interest in climate change.

★   ★   ★

The 1980s was the decade when many Americans bought their first home computer, learned to heat up coffee in a microwave, and became aware of ominous phrases like the "greenhouse effect" and "global warming." Over the course of the decade, mainstream science writers, national politicians, and, eventually, leading news magazines pushed the issue into prominence.

The greenhouse effect was also shoved into the spotlight by the federal government. In 1983, soon after I became a staff engineer there, the Environmental Protection Agency echoed a recent report about the potentially catastrophic impact of climate change. Their announcement, which was reported on by the *New York Times*, marked the first time that any government agency had declared climate change to be not just a potential threat but urgent and real.

The following year, the handsome, young Tennessee congressman Al Gore organized Congressional hearings on climate change. Among the list of speakers was Carl Sagan, the host of *Cosmos*, public television's most-watched series. An astronomer, Sagan was not as qualified to present on climate issues as some of the other panelists who spoke. But Gore wasn't just trying to educate the public in the details of the greenhouse effect, he wanted attention for the issue. Only Sagan could attract the maximum number of television cameras.

The weather throughout the 1980s also cooperated in raising concern. By mid-decade, 1981 and 1983 were the world's two hottest years on record. After several more chart-topping years, the decade itself was also declared the hottest ever. The rising mercury also made it possible for the media to point to a tangible, everyday result of carbon buildup. As the decade continued, new computer technologies allowed for more advanced climate modeling. Sophisticated research data from all points of the earth continually poured in. The threat was taking shape.

By the late 1980s, climate concerns had hit the big time. Following a scorching summer, the October 19, 1987, cover of *Time* portrayed a radiating earth encased in a glass house and the boldface headline "The Heat Is On: How the Earth's Climate Is Changing."[17]

As *Time*, *Newsweek*, national television news, and other outlets gave climate science a higher profile, the issue also attracted widespread public concern. In a 1981 poll, about a third of respondents said they were aware of or had read about the greenhouse effect. In a separate survey from the following year, 12 percent of Americans said they considered the greenhouse effect to be a serious problem. By 1989, these numbers had increased dramatically. When researchers repeated the question from the 1982 poll, they now found 41 percent of respondents considered the greenhouse effect to be serious, a nearly thirty percentage point rise in concern in seven years.[18]

But even though awareness and concern about climate change had spread well beyond a relatively small group of scientists, many Americans still had no idea how the science behind this new threat worked. Large numbers of people, for example, suspected that nuclear tests or the space program were the primary culprits behind what was actually a greenhouse gas–driven process. There were, and still are, a number of major barriers to grasping the nature of this environmental threat.

For one, global warming is the result of a novel form of environmental disaster. A poisoned fish floating by in a river or a city covered by a cloud of smog or a barge loaded with garbage that can't find a port points to a clear and immediate pollution hazard. The buildup of greenhouse gases takes place over many decades, and most of the gases can't be seen or smelled. While the public had proven willing to act on environmental issues that posed an imminent and tangible threat to their communities, action on climate change required unwavering confidence in scientific data. Accepting that human activity could change something as massive as the temperature of the oceans was another major conceptual hurdle.

Second, the public's initial education on climate change came from scientists, who are by nature unwilling to rely on snappy, television-ready sound bites. They are trained to explain the theory with all its uncertainties and caveats included—the stuff of documentaries, not primetime news. Additionally, mainstream media might cover the most recent climate change news without delving deeper into the science. This is unfortunate because, by the mid-1980s, there were a number of great opportunities to explain the relationship between carbon emissions and a warmer planet.

One such illustrative research effort took place in the Antarctic summer of 1983–84 at Siple Station, a research complex located near the magnetic South Pole. Operating in frigid temperatures, teams from the University of Nebraska and University of Bern took a coring drill—an empty tube with a ring of teeth around one end—and gradually cut 200 meters, or over 650 feet, into the ice. As they pulled the cylindrical pieces up, one frozen segment at a time, the team was also withdrawing an extremely well-preserved record of climates past. In this particular drilling, each meter of ice equaled roughly one calendar year. Up to 144 meters, when the research teams removed data that had sat frozen since about 1834, they could date the ice to an accuracy of about plus or minus two years.[19]

The icy columns also retained other information too. The chemistry of the frozen water recorded annual temperatures. Tiny bubbles of air trapped in the ice contained fragments of previous atmospheres. By analyzing the chemical makeup of the gases in the bubbles, the scientists could recreate accurate historic measurements of carbon dioxide levels going back a century and a half. The results were dramatic: the concentration of carbon dioxide had been increasing rapidly throughout the period and, right along with it, the earth's surface temperature. The core provided rock-solid evidence for Arrhenius's climate change theories that, themselves, dated back almost eighty years.

Since then, researchers have drilled ever deeper into ice sheets and even further back in time. Ten years after the Siple Station dig, a team from the European Institute for Ice Coring (EPICA), housed in a sleek, shiny barnlike building in Antarctica, began drilling into a glacier called Dome C. Eventually, the expedition was able to retrieve data from nearly 800,000 years ago. This massive core put the findings of Siple in a huge—and stark—context.[20]

The Siple dig gave climatic information about 150 years back in time, a time before there were reliable records for temperature and carbon dioxide concentrations. The EPICA effort provided frozen data points so far back in geologic time that scientists could track the rise and fall of temperature and carbon dioxide over the course of multiple ice ages. Over the previous eight hundred thousand years the two had consistently risen and fallen together within a certain operating range. Nearly one hundred years after Arrhenius had laid out his "hot-house theory," scientists could supply hundreds of centuries of climate data to back it up. Just as the Swedish scientist had predicted, carbon dioxide and temperature volleyed back and forth nearly in unison. This relationship between carbon dioxide and earth's temperature is the fundamental mechanism of climate change, and one that has yet to be effectively explained to the American public. But research at places like EPICA also added a great deal more detail to the picture.

★   ★   ★

As simple, almost graceful, as the rise and fall of carbon dioxide and temperature are, they don't take place in a vacuum. The earth is the very complex nexus of multiple systems that are constantly interacting with the basic mechanics of global warming. The changes initiated by a slight rise in temperatures set off additional dynamics that can either accelerate or decrease the heating of the atmosphere.

For example, water vapor is a powerful greenhouse gas. Air that is warmer can also hold more water vapor. So, as the planet heats up, the atmosphere will retain more water vapor. This additional water vapor in the atmosphere retains more heat, further increasing temperatures. At the higher heat, the cycle repeats itself, with the hotter air able to retain even more water vapor, which then traps more heat, and so on. This dynamic is called a "positive feedback loop," because the two elements, increased temperature and additional water vapor, act like magnifiers of each other. An increase in one factor drives a further rise in the other.

Other parts of the ecosystem also act together in positive feedback loops. For example, the large white sheets of ice in the Arctic Ocean currently reflect a substantial amount of sunlight back into space. But as temperatures rise, the ice will melt into the ocean. The darker-colored ocean water absorbs more sunlight, increasing the temperature of the ocean. The warmer water then melts more ice, continuing the positive feedback loop.

Even though they absorb more light than ice, the oceans do play an important role in mitigating the amount of carbon dioxide in the atmosphere. Back in the 1940s, Roger Revelle knew that the oceans sucked up lots of the excess carbon dioxide created by human activities. Although most of the carbon dioxide absorbed by the seas bubbles back out of the water like soda pop in a glass, the water still retains roughly a quarter of the carbon dioxide that humans produce. In fact, the oceans are the largest carbon sinks on the planet. But, as the planet heats up, they become less reliable sponges for the greenhouse gas.

It turns out that colder water is much better at absorbing carbon dioxide than warmer water. As a result, nearly half of this absorption process happens in the Southern Ocean around Antarctica, a fairly small portion of all the world's salt water.[21] But as the temperature of the ocean increases, water everywhere will become less effective at absorbing carbon, increasing carbon dioxide levels in the atmosphere that will, in turn, prompt a further rise in temperature.

As fascinating and important as these numerous feedback loops are to understanding the broader dynamics of climate change, the phenomenon's basic metrics and warning signs are still temperature and the levels of atmospheric carbon dioxide. The ancient EPICA core sample revealed that, as ice ages came and went roughly every 100,000 years, average Antarctic temperatures rose and fell from about −8 to 4 degrees Celsius (18 to 39 degrees Fahrenheit).[22] The coldest of these extreme temperature fluctuations reflected an earth in which massive glaciers moved as far south as the San Bernardino Mountains just east of Los Angeles.[23] In the warmest eras, glaciers retreated back to the poles and sea levels rose by hundreds of feet. During these periods, the level of carbon dioxide in the atmosphere also rose and fell correspondingly. The ice-age low for carbon dioxide in the atmosphere was around 180 to 185 parts per million (ppm). As the temperatures peaked, carbon dioxide rebounded upward, maxing out at around 285 to 300 parts per million. Over eight hundred millennia, temperatures and carbon dioxide rose and fell within this range.[24]

Then, about two hundred years ago, the numbers measuring carbon dioxide concentrations began going crazy, rising at a rate that was unprecedented over the previous 800,000 years. In 1800, carbon dioxide measured around 300 ppm, a level that marked the high end of concentrations established over the previous eight hundred millennia. Today the concentration of atmospheric carbon dioxide sits near 400 ppm. This is not only an alarming deviation from the norm. This stratospheric rise in carbon dioxide levels closely mirrors the beginning of the Industrial Revolution, precisely when humans began burning carbon-intensive materials at unprecedented rates.[25]

Evidence from coral reefs, tree rings, and ice cores tells us that past climate changes caused huge losses of life in every ecosystem on earth. But these destructive transitions also took place over many millennia, giving life on land and in the oceans time to adapt. The current massive buildup of greenhouse gases in the atmosphere, due largely to the burning of fossil fuels, dates back only to the late nineteenth century. The impacts of this rapid change on the climate are already beginning to be felt and will accelerate beyond this century. Without doubt, this unprecedented pace of change threatens the entire food chain from the bottom up.

By near the end of the twentieth century, climate scientists knew with increasing certainty that human-caused global warming was happening. They also knew that the impacts of extreme climatic shifts for life on earth would not be pretty, to say the least. But it would take a great deal more persuasion before people would begin to accept the need to move away from a fossil fuel–based economy. Communicating these ideas to a mainstream audience would not be easy. But, toward the end of the decade, one scientist would sound a clarion call for action on an undeniable threat.

★   ★   ★

June 24, 1988, was a particularly hot day in Washington, DC, making it perfect for a Congressional hearing on climate change. The previous night, staffers had strategically opened all the windows in the hearing room, letting in the muggy night air. The air conditioning labored to keep the room comfortable as testimony began. It was in this setting that a NASA scientist with a red tie and tan blazer sat down to give his extraordinary testimony. Going well beyond what any other scientist had been willing to state, James Hansen unequivocally declared to some of America's top elected officials that this sweltering heat was only going to get worse.

Surrounded on either side by other climate experts, James Hansen spoke unreservedly into the two microphones facing him. For twenty minutes, Hansen made three main points: the earth is warmer in 1988 than any time in the history of instrumental measurements; current levels of greenhouse gas concentrations are so high that they can be directly correlated to the "greenhouse effect"; and recent computer climate simulations estimate that the greenhouse effect is large enough to trigger repeated summer heat waves, among other extreme events.

To his left, an overhead projector showed numerous charts and data bolstering the overwhelming scientific consensus that human activity, particularly burning fossil fuels, had begun to change the earth's climate and weather patterns. In the coming decades, Hansen predicted, there would be a dramatic increase in droughts and heat waves, particularly in the southern and midwestern parts of the country. Global warming, he concluded, is a lethal phenomenon that poses a real and serious threat to the future of the planet.

As if to underscore Hansen's words, that summer went on to be the hottest in recorded history, and a prolonged drought engulfed much of country, causing massive forest fires, killing between 5,000 and 10,000 people, and causing $70 billion in agricultural losses in the United States. For some areas, the drought didn't end until September of the following year.[26]

As he left the hearing, Hansen fired off one more cry for action to the assembled reporters. "It's time to stop waffling so much and say that the evidence is pretty strong that the greenhouse effect is here."[27]

The officials in Congress weren't convinced. They went on to spend over $5 billion in drought assistance to farmers but failed to pass any piece of legislation that addressed the greenhouse gases piling up in the atmosphere.[28] Despite the more than 150 years in which modern global warming overcame skepticism from the scientific establishment, and in the face of conclusive evidence documented by tens of thousands of scientists worldwide, they still haven't acted. Amazingly, a relatively small group of people and interests have managed to keep the United States government from ever targeting greenhouse gas reduction.

# 3

# HOT AIR

By the end of 1980s, climate change not only had the American public's attention but was prodding the federal government toward action. In 1988, a few months after James Hansen's unprecedented Congressional testimony, the United Nations created the Intergovernmental Panel on Climate Change (IPCC), a loose confederation of more than 2,000 scientists from around the world. With so much changing, contradictory, and outright false science surrounding what was clearly a serious global issue, the panel would be the final scientific authority on global environmental issues. The IPCC wouldn't conduct any original research but rather would review the reams of published literature on climate change. The group's first task was to examine the current status of global warming and its potential implications for policymakers.

The next year, the global movement for climate action got another huge push when UK prime minister Margaret Thatcher stood before the annual meeting of the United Nations General Assembly, outlined the dangers of soaring carbon emissions and called for the ratification of an international treaty.

"We are seeing a vast increase in the amount of carbon dioxide reaching the atmosphere," declared Thatcher. "The result is that change in the future is likely to be more fundamental and more widespread than anything we have known hitherto. Change to the sea around us, change to the atmosphere above, leading in turn to change in the world's climate, which could alter the way we live in the most fundamental way of all."[1]

Thatcher, who held an undergraduate degree in chemistry from Oxford University, was a much more high profile spokesperson for climate change than James Hansen. She was also the first world leader to call for action on the problem. Perhaps most important, Thatcher was an icon of the conservative deregulatory politics identified with President Reagan in the United States. The Iron Lady had unassailable free market credentials. Though she was reviled by the British left-of-center parties, Thatcher could also not be dismissed by critics of regulatory action as "a communist" for whom global warming was simply the means to impose government control.

Even President Ronald Reagan had asked for increases in the research budgets for the US Global Change Research Program. A 1988 National Academy of Science report called *Our Changing Planet: A US Strategy for Global Change Research* recommended $134 million for fiscal year 1989 and $190 million in 1990.[2] The Reagan administration was no supporter of action on climate change. But, even if the report was largely a cynical stalling move, its inclusion in the budget of a conservative antiregulatory Republican was still an impressive reflection of global warming's growing political clout.

In 1990, the first report of the Intergovernmental Panel on Climate Change unambiguously concluded that human activity had been "inadvertently changing the climate of the globe through the enhanced greenhouse effect by $CO_2$ emissions." It also confirmed that the average temperature of the planet had risen within the past hundred years and that the current inaction on greenhouse gas emissions was digging humanity an ever deeper hole. Political momentum finally seemed to be tipping in climate science's favor at the highest levels of the US and other national governments, but the battle had only just begun.

★   ★   ★

In late January of 1990, a policy debate over action on climate change raged within the administration of President George H. W. Bush.[3] The controversy surrounded an address President Bush was presenting to the Intergovernmental Panel on Climate Change on February 5. Bush, who had only been in office for one year, was expected to describe just how Americans' new environmental concerns would translate into political action. Officials at the Department of Energy and the State Department urged him to call for strong measures to curb greenhouse emissions, but Bush had previously hedged on the issue. EPA administrator Bill Reilly was pushing in this direction too. I admired his willingness to speak out on important environmental issues. Later that year he played an important role in securing the 1990 amendments to the Clean Air Act. Unfortunately, the president was also listening carefully to the advice of his chief of staff, John Sununu, who favored a more conservative line.

The speech that finally emerged could be described as a compromise. Bush declared that "The United States remains committed to aggressive and very thoughtful action on environmental issues."[4] But, while paying lip service to climate change, Bush disappointed environmentalists by not offering any new recommendations for specific action to slow or reduce global warming. The issue was too big to ignore and too real to refute, so the president decided to stall.

After the speech, Chief of Staff Sununu, who claimed to have heavily edited the final version of Bush's speech, told reporters that "faceless bureaucrats" wanted to cut off "our use of coal, oil, and natural gas."[5] He also said that regulatory action would make it too expensive or otherwise difficult for Americans to drive their cars. But, in addition to making inflammatory and unfounded statements about action to curb global warming, Sununu was one of the primary advocates in the Bush administration of a more nuanced and well-organized attack on climate science that had been honed over previous decades.

In interviews, Sununu stressed the indeterminate aspects of research on global warming, claiming that not enough was known to take strong—or, as he called it, "radical"—action. With a degree in nuclear engineering, Sununu impressed Bush with his ability to explain the technical issues behind climate science. Sununu ran computer models that he claimed showed uncertainties clouding the understanding of global warming. In his education of the president, Sununu constantly stressed the ambiguous aspects inherent in any developing field of study instead of the near certainties of global warming's destructive impact.

This ongoing effort to undermine the credibility of climate science within the administration became a public embarrassment in April of 1992. At an international environmental conference, a Bush administration document with a list of talking points for American delegates was inadvertently leaked to the other countries' representatives. The memo advised Americans to avoid discussions about whether there is or isn't global warming. "In the eyes of the public, we will lose this debate. A better approach is to raise the many uncertainties that need to be better understood on this issue."[6] In damage control mode, the Bush administration quickly back-stepped and distanced itself from the document. But the technique of obfuscation remained a central weapon in their crusade to prevent action on climate change. Opponents of regulatory efforts thought they had found the Achilles' heel of the surging movement to take serious action on global warming. It was not the first time that a campaign of obfuscation and uncertainty had been deployed to delay regulation.

★   ★   ★

In 1969, marketing executives at the tobacco company Brown & Williamson, manufacturers of over a dozen cigarette brands including Kool, circulated a document they thought could help them "counter the anti-cigarette forces."[7] At the time, all tobacco companies were facing medical evidence that cigarettes cause cancer. Because Brown

& Williamson couldn't disprove these claims with better science, the company's solution was to create uncertainty about their products' health risks. In a now infamous passage, the document said, "Doubt is our product, since it is the best means of competing with the 'body of fact' [linking smoking with disease] that exists in the mind of the general public. It is also the means of establishing a controversy . . . if we are successful in establishing a controversy at the public level, there is an opportunity to put across the real facts about smoking and health."

A list of objectives followed, the first of which stated that the company had to "set aside in the minds of millions the false conviction that cigarette smoking causes lung cancer and other diseases; a conviction based on fanatical assumptions, fallacious rumors, unsupported claims and the unscientific statements and conjectures of publicity-seeking opportunists." In other words, by cynically creating confusion, the marketing executives sought to hold back regulatory action. The cigarette companies' tactics were very successful at pushing off regulatory action for several decades. Opponents of action on global warming wanted to recreate this success.

By the mid-1970s, a few years after the Brown & Williamson document was released, a combination of free market ideologues and business interests had joined forces to combat the EPA and a newly empowered environmentalism. The heads of multiple industries issued dire predictions that with new regulations their whole business model might become unviable. But, realizing that the public would be more likely to believe these antiregulatory arguments when delivered by seemingly disinterested third parties, the companies took another page out of the tobacco industry playbook.

Tobacco companies had funded and created front groups to spread complicated technical arguments and conduct sophisticated public relations campaigns designed to confuse the facts supporting any government regulation. The fossil fuels industry began supporting groups that targeted any environmental legislative or regulatory action, including on the nascent issue of global warming. The organizations developed studies and advertisements, but their most successful tactic was promoting the work of the increasingly rare scientists willing to speak out against the near consensus on global warming. These scientists could use their credentials or previous reputations to soothe the public into believing that there was nothing to worry about.

The book *Carbon Dioxide: Friend or Foe?* by a respected scientist named Sherwood Idso was a perfect example. In the book, Idso claims that increased greenhouse gas emissions would actually increase food yield and provide other benefits.[8] His positive take on global warming was, in fact, true for some parts of the planet for short periods of time. But, overall, Idso's arguments gave a false impression of the future impacts that climate change would have on agriculture. It was also popular and controversial, exactly what industry interests wanted.

In 1989, at the end of a decade in which surging public support made action on global warming seem likely, the fossil fuels industry created a group called the Global Climate Coalition. The companies funding the group, including Exxon, Shell, British Petroleum, Ford, and Chrysler, specifically wanted to counter the impact of the United Nation's newly created Intergovernmental Panel on Climate Change. They had watched with alarm as science and public concern began steering governments toward action to curb the emission of greenhouse gases. These carbon-intensive companies saw such measures as a serious threat to their businesses. They decided the best way to kill or at least stave off legislative or regulatory action was to target the science of climate change. Their front group's name was intended to lend the industry-funded organization a veneer of impartiality. But from their Washington, DC–based offices, the Global Climate Coalition conducted a systematic and well-funded campaign to distort and obfuscate climate science's consensus on the threat of global warming.

In the early 1990s, for example, the group distributed a fact sheet to lawmakers and journalists claiming, "The role of greenhouse gases in climate change is not well understood." While there was some truth to their statement that "scientists differ" on climate change–related issues, the basics of the greenhouse effect had been largely accepted by most climate scientists for decades. And any credibility the group may have had as an advocate for open debate was shattered when a document came to light revealing that the Global Climate Coalition's own scientists and technical experts had advised their bosses that the greenhouse effect and impact of human emissions of carbon dioxide could no longer be denied. Nonetheless, the Global Climate Coalition continued to work to undermine any action on the issues, pushing this discredited line for years longer.[9]

Although the forces opposing action on climate change certainly included ideologues, the efforts of front groups like the Global Climate Coalition to blur climate science were very pragmatic. These organizations initially argued that the world was not actually warming, that any measurements were flawed. Once that claim had been largely discredited, the groups took the line that the earth was heating up, but that this warming was merely a natural cycle of the earth and not caused by human activities. As these arguments too were overwhelmed with contrary evidence, they instead argued that any changes from the human-driven climate change would be tiny and harmless. The anti–climate science groups didn't care about winning a debate; they wanted to stall government action as long as possible. In this sense, they were scoring lots of victories.

Even as public awareness of global warming dramatically increased in the 1980s, public understanding of the science behind the issue remained unclear, and Americans' concern about global warming had not translated to federal action. Having taken a page out of the cigarette industry's playbook, the opponents of climate change science had succeeded in sowing confusion and forestalling action. And the two antiregulatory campaigns would become even more sophisticated and intertwined in the coming decade.

★   ★   ★

In 1993, I was the director of the EPA's Office of Radiation and Indoor Air, a job that gave me a front row seat to the misinformation tactics employed by the cigarette companies. That year, the tobacco industry

tried to undermine a landmark study my team had released concluding that secondhand smoke was responsible for three thousand lung
cancer deaths among nonsmokers. I knew we had very strong technical support for our report. Bob Axelrad, my senior indoor-air expert,
had worked for several years with other EPA scientists to determine
the links between secondhand smoke and cancer. There was also solid
science linking tens of thousand of cases of bronchitis and pneumonia
in children to what was sometimes called passive smoking. The report
made public smoking bans across the country very likely, setting off a
panic within the tobacco industry. Within a few months, the world's
largest tobacco company, Philip Morris, formulated a response.

In February 1993, Ellen Merlo, senior vice president of corporate
affairs, sent a letter to Philip Morris CEO William Campbell. "Our
overriding objective is to discredit the EPA report," wrote Merlo.
"Concurrently, it is our objective to prevent states and cities, as well as
businesses, from passive-smoking bans."[10]

Finally, Merlo and Philip Morris's public relations company, APCO,
recognized that any messages undermining the negative health effects
of smoking that could be traced back to the company would be dele-
gitimized. They decided to create not just a 1970s-style front group to
distribute their message but a next-generation front group that even
further distanced itself from Philip Morris. APCO became a pioneer
of "astroturfing," creating movements and groups that appeared to be
organic grassroots actions by concerned citizens. To avoid "cynical"
big name newspaper and media, APCO also chose secondary markets
to launch efforts and engaged almost exclusively with local media.

A year later, APCO unveiled the flagship organization, The
Advancement of Sound Science Coalition, or TASSC. The public
relations firm further buried the corporate money trail back to Philip
Morris by encouraging TASSC to question the science surrounding a diverse range of topics instead of focusing narrowly, and thus
somewhat obviously, on tobacco. TASSC might claim that secondhand smoke, cell phones, and pesticides were all subject to unfounded
health concerns and regulation. Or, under the antiregulatory flag,

the group might question scientific studies not just about smoking but also about nuclear waste disposal, biotechnology, and . . . global warming.

TASSC took this work seriously. Over the course of the 1990s, the front group became so effective at labeling the evidence on global warming "junk science" that it eventually received some funding from Exxon, a fossil-fuel company with no deep interest in the regulation of secondhand smoke. In 1997, a Washington, DC, veteran of corporate funded antiregulatory think tanks named Steven Milloy took over the presidency of TASSC. His website, junkscience.com, became so popular that he later became a science correspondent for FOX News in 2002, providing him with an enormous audience to debunk health and environmental concerns like global warming.

The tobacco industry also launched a less subtle, direct attack on the EPA by suing us, prompting the court to freeze EPA's study. For seven years, EPA scientists fought back until, in 2002, the court finally threw out the cigarette industry's challenge. Unfortunately, this wasted time in court cost lives as people continued to die from the cumulative effects of secondhand smoke.

Science did score some victories against the industry front groups during this period. In 2001, for example, the Global Climate Coalition crumbled following an Intergovernmental Panel on Climate Change report on the extreme dangers of global warming. Member companies decided that they no longer wanted to be associated with a group that was prominently identified with climate change denial. Nonetheless, these campaigns to discredit climate science continue to influence public opinion to this day. The seemingly innocent question, "Do you believe in climate change?" is one marker of their success. Such a statement reduces the scientific reality of climate change to a choice. As a Union of Concerned Scientists senior climate expert, Brenda Ekwurzel, pointed out to me during our interview, the basic mechanics and effects of climate change are so well documented that informed supporters should say, "I don't *believe* in climate change—I know it is true."[11] By positing that there are two plausible sides to a theory that is supported by 98 percent of scientists world-wide, climate change deniers have succeeded in demoting the issue to one in which reasonable people can choose to disagree. In this world, the skeptics have already won.

★  ★  ★

In 1993, Bill Clinton was sworn in as president. It was the first time in nearly twelve years that a Democrat had sat in the Oval Office. Both houses of Congress had Democratic majorities. Best of all, the new vice president was Al Gore, who, as a senator, had made his name as an early advocate for action on climate change. After the obstructionism on climate issues of the Reagan and Bush presidencies, environmentalists—including me—were euphoric.

Over the past four years, proponents of climate change action had watched the country stall and balk at every turn. At a landmark environmental conference in Rio the previous year, the Bush administration had fought European nations' efforts to take the warnings of the Intergovernmental Panel on Climate Change seriously. The eventual result, an agreement to lower emissions to the 1990 level by the year 2000, included no binding penalties.

At nearly the same time, Clinton had been on the campaign trail promising action to mitigate the damaging effects of greenhouse gases. In a speech on Earth Day 1992 he said, "Our addiction to fossil fuels is wrapping the earth in a deadly shroud of greenhouse gases." The contrast between the incoming and outgoing administrations was stark. That same year, the Union of Concerned Scientists sponsored a forum called World Scientists' Warning to Humanity that released a document calling for the "fundamental change" necessary to address a range of security and environmental issues. The call to action was signed by 1,700 scientists, including numerous Nobel Prize winners. The long-delayed action on global warming seemed almost certain, but Clinton's environmental efforts would be resisted by legislators from both parties and fiercely opposed by well-funded and increasingly sophisticated campaigns to discredit climate science.

In 1993, the Clinton administration attempted to make good on its promises by pioneering two efforts to tackle climate change. In February, Clinton announced a British Thermal Unit (BTU) tax, a fee on the heat content of energy, claiming it was, "the best way to provide us with revenue to lower the deficit, because the tax also combats pollution, promotes energy efficiency, and promotes the independence economically of the country as well as helping to reduce the debt."[12]

On Earth Day, a few months later, Clinton reversed the government's previous position on climate change and promised to reduce greenhouse gas emissions to their 1990s level by the year 2000. This brought the United States into line with the agreement accepted by most developed nations at the Rio conference.

Unfortunately, this was the high point of Clinton's first term. His BTU tax hit staunch resistance from both parties in Congress and was eventually weakened to the point of uselessness. When President Clinton revealed the details of his Climate Change Action Plan, they were a series of voluntary actions suggested for industry, including improving energy efficiency and promoting green technologies. The president did have some successes, including a regulation that cut back emissions of methane, a greenhouse gas, from landfills. In addition, voluntary energy reduction programs were introduced to promote reductions in carbon dioxide around the country. However, many environmental groups panned his action plan as completely inadequate to the enormous task at hand.

Meanwhile, climate science continued to demand action. In 1995, the Intergovernmental Panel on Climate Change issued its second assessment report, indicating that greenhouse gas concentrations from human activity were continuing to climb upward in both the United States and most other developed nations. One metric that got the public's attention was the confirmed heating of the planet's surface. Nineteen ninety-five was the hottest year in historical records going back to 1866. But that same year also saw a newly elected Republican

House and Senate obstructing all administration priorities, including efforts to save the environment.

Nonetheless, after Clinton's reelection in 1996, he and Gore hoped to get back on track. In early 1997, Clinton established the White House Climate Change Task Force to direct the administration policy and educate the public on the dangers of increased greenhouse gas emissions. The task force reached out to public and private stakeholders as well as environmental, labor, and industry groups. It also transcended international boundaries, shuttling between nations to discuss the US position on climate change with world leaders.

The single most important global warming event of the late 1990s was the Kyoto Protocol, a binding international treaty based on the 1992 Rio framework. The protocol called for countries to agree to firm reduction targets for their national greenhouse gas emissions. In most nations, that meant rolling emissions back to 1990 levels. The administration was supportive of the measures and even engaged in drafting the protocol's documents. But all international treaties have to be approved by the Senate, which proved a hopeless task.

In response to the Kyoto Protocol, a web of conservative think tanks, big oil companies, trade associations, and energy industry insiders hatched a plan to derail Clinton administration policy efforts on global warming by using now familiar tactics. Often meeting at the American Petroleum Institute's headquarters in Washington, DC, the network of global warming opponents proposed to spend millions of dollars to recruit a host of like-minded scientists to convince the public, media, and lawmakers that the climate science embraced by the Clinton administration was built on faulty evidence and loose foundations.

The eight-page memo spelling out how the group would scuttle the Kyoto treaty was not innovative. "Because the science underpinning the global climate change theory has not been challenged effectively in the media or through other vehicles reaching the American public," the document states, "there is widespread ignorance, which works in favor of the Kyoto treaty and against the best interests of the United States."[13] The head of the group, Joe Walker, asserted that, when average citizens, the media, and industry leadership "understand uncertainties in climate science; recognition of uncertainties becomes part of the conventional wisdom," and then victory has been achieved.[14]

The tactics masterminded by the tobacco industry four decades earlier lived on in efforts to scuttle action on global warming. Instead of "doubt is our product," the rallying cry for anti-climate change crowd was something like "uncertainty is our ally." But even though the campaign to sow seeds of doubt about the science of global warming was hardly fresh, it was still remarkably successful.

In 1997, before President Clinton had even signed the landmark Kyoto Protocol that created binding greenhouse gas reduction targets for the signatory countries, the Senate passed a bipartisan resolution preemptively voting down the treaty. "There's no way we'll even be close in the Senate to ratifying this agreement. We will kill this if the president signs it," said Senator Chuck Hagel, one of the resolution's sponsors.

The Senate approved the anti–Kyoto Protocol resolution unanimously, 95-0. President Clinton signed the climate treaty in 1998 but never submitted it to the Senate. But even though the strongest international agreement yet created had no chance of becoming binding US law, industry opponents weren't done with their efforts to scuttle the accord.

★   ★   ★

In 2001, the third assessment by the Intergovernmental Panel on Climate Change established a cautiously phrased scientific consensus. It was much more likely than not, said the report, that human civilization faced severe global warming as a result of "dangerous anthropogenic interference with the climate system."[15] Although estimates about sea level rise, droughts, and temperature increases would be refined as the new century proceeded, the major factor in future climate change was human policy.

The IPCC's conclusions were reviewed and endorsed by leading scientific organizations, like the United States' National Research Council, as well as the national science academies of more than a dozen countries, including most of Western Europe, as well as China, India, Indonesia, and Brazil. By the first year of the new millennium, global temperature had been rising substantially for two decades. The Arctic had lost noticeably large swaths of ice, and extreme weather events associated with global warming were more common. The rest of the developed world was pursuing serious global action on climate change, but it needed American participation and leadership. That same year, George W. Bush took office.

Environmentalists would certainly have preferred the Democratic candidate, Al Gore, in the White House. But there was some hope that Bush might steer at least some bipartisan progress on environmental issues. Although he had spent years in the oil business, as governor of Texas Bush had encouraged the use of wind energy and eventually made the state the leading producer of wind-generated electricity in the United States. During the 2000 campaign, he had advocated reducing carbon dioxide emissions from power generation.[16] Nevertheless, his administration's environmental and energy policies soon seemed to be dictated by fossil fuel companies. As reports set forth the dangers of global warming with greater and greater certainty, the Bush administration resorted to tactics clearly intended to avoid science-based discussion on the issues.

The first victim of this strategy was Bush's newly-appointed EPA administrator, Christine Todd Whitman, the former governor of New Jersey. In March of 2001, Whitman attended an international global warming summit in the northern Italian city of Trieste. Outside the meeting, thousands of anti-globalization protestors marched in the rain, chanting and hurling colored smoke bombs at riot police. Inside, the tall, earnest Whitman spent two days trying to convince the leaders of the other major industrialized nations that the United States was serious about attacking the problem of climate change. Since Bush had already criticized the Kyoto Protocol, Whitman had a tough sell.

"The president has said global climate change is the greatest environmental challenge that we face, and that we must recognize that and take steps to move forward," Whitman told her counterparts.[17] To accomplish this, she added, the Bush administration would press for bipartisan legislation to limit emissions of carbon dioxide.

After Trieste, CNN ran a headline, "WILL BUSH TURN GREEN?" Environmentalists, myself included, certainly hoped so. Whitman's speech was in part an attempt to buy time for the Bush administration to figure out what exactly it was going to do on climate change. But her promise was not a ruse. Before going to Trieste, Whitman had run her speech by White House chief of staff Andy Card as well as National Security Advisor Condoleezza Rice. Neither had offered any objections. But as fossil fuel companies tightened their clamp on the Bush administration's global warming policy, the ground shifted beneath Whitman's feet.

Once back in DC, Whitman began to hear through back channels that the White House was moving away from the promises she had just delivered to the G8 nations. A few days later, she went to a meeting in the Oval Office to convince the president and his chief of staff, Andy Card, that they shouldn't abandon a policy of setting limits on carbon emissions. As she was leaving, Whitman remembers seeing Vice President Cheney walking toward the office as he was putting on his coat.

"Well, was the letter ready?" he asked.[18]

Whitman had no idea at that moment, but the letter Cheney referred to was the result of pressure from legislators, lobbyists in the coal and oil industries, and the vice president himself. It officially reversed Bush's promises on carbon, killed the United States' international credibility, threw Whitman and the EPA under the bus, and set the stage for a White House energy policy largely driven by coal and oil companies.

Bush went on to withdraw the United States from its purely symbolic support of the Kyoto Protocol. Soon after announcing the reversal, the head of the US delegation to the Kyoto negotiations met with the oil lobbyists from the Global Climate Coalition. He thanked them for their expertise and said that President Bush had "rejected Kyoto in part based on input from you."

That same year, 2001, saw many of the corporate members of the Global Climate Coalition decide that the science on global warming was too overwhelming to be associated with the front group. The Bush administration's climate policy was so extreme that it was taking advice from an organization whose members were leaving it in droves. By 2002, the Global Climate Coalition was defunct. But their policies lived on under the Bush presidency.

★ ★ ★

One of the most flagrant indicators of this willingness to work closely with the fossil fuel industry was the appointment of Philip A. Cooney as the chief of staff of the president's Council on Environmental Quality. Cooney had spent the previous fifteen years as an oil industry lobbyist for the American Petroleum Institute, leading their climate policy team. Unsurprisingly, his job had primarily been killing and weakening any regulations on greenhouse gas emissions.

Now ensconced in the administration, Cooney got busy doing more or less the same work he had at the American Petroleum Institute, fomenting uncertainty about climate change science. In his position, Cooney was able to review and edit government scientists' reports on global warming. A *New York Times* article cited several instances in which Cooney, who had no scientific training, tweaked reports to minimize the links between greenhouse gas emissions and global warming. In many cases, Cooney's edits appeared in the reports' final versions. His work—for example, adding a phrase like "significant and fundamental" before the word "uncertainties"—was right out of the fossil fuels industry playbook.[19]

The *New York Times* cites an October 2002 draft of a summary of government climate research called *Our Changing Planet*, where Cooney added the word "extremely" to this sentence: "The attribution of the causes of biological and ecological changes to climate change or variability is extremely difficult." Cooney also eliminated text that provided evidence of global warming's impact, including a section that contained projections on the reduction of mountain glaciers and snowpack. In the margins he dismissed the scientists' work with the comment "*straying* from research strategy into speculative findings/musings." In another case he changed the meaning of the sentence "Many scientific observations indicate that the Earth is undergoing a period of relatively rapid change" by rewriting it as "Many scientific observations point to the conclusion that the Earth may be undergoing a period of relatively rapid change." Cooney resigned in 2005 following public revelation about his numerous edits to government scientists' climate change work. Soon thereafter, he took a job at Exxon, but the silencing of climate change science continued.

★   ★   ★

As the evidence of global warming became harder to blur, bury, or question, officials in the Bush administration used a less subtle technique: muzzling government climate scientists. Their highest profile target was James Hansen, the NASA scientist who in 1988 demanded that the government stop waffling on the greenhouse effect. On December 6, 2005, Hansen gave a speech at the annual meeting of the American Geophysical Union in San Francisco, during which he called for prompt reductions in emissions of the greenhouse gases linked to global warming. He warned that without America taking the lead, climate change would turn the earth into a "different planet." On December 15, Hansen, Director of the Goddard Institute for Space Studies, issued findings that 2005 was likely the hottest year in over a century.

The following month, January 2006, Hansen claimed in an email to the *New York Times* that NASA leadership had begun to stop him from speaking out. Top officials began phoning his New York–based institute warning of "dire consequences" if Hansen kept issuing his reports.[20] Bush political appointees then imposed restrictions intended to silence him, which included having supervisors replace him in interviews and rejecting an NPR request for an interview, reportedly because the news outlet was too liberal. Hansen was not the only scientist who claimed to have been subjected to such efforts.

Scientists at the National Oceanic and Atmospheric Administration and Bureau of Land Management said they too had been subject to measures designed to quiet climate research that didn't toe the Bush administration line. Outside reporters covering the issues, including the *New York Times*'s Andrew Revkin, supported such claims, saying that the government climate scientists he used to interview five years previously could now take calls only if the interview was pre-approved by Bush administration officials and actively monitored by a public affairs officer.

★   ★   ★

In 2007, The United Nations' Intergovernmental Panel on Climate Change released its fourth report, stating that warming of the climate system was "unequivocal" that "unmitigated climate change would, in the long term, be likely to exceed the capacity of natural, managed, and human systems to adapt" and that many of the impacts could still be "reduced, delayed or avoided by mitigation." In other words, the threat was real, the potential to destroy humanity was confirmed, and the time to act was now.

The following year, Barack Obama was elected president. During his campaign he had promised to take action on climate change. Millions of Americans hoped he would finally succeed, none of them more than my team at the Environmental Protection Agency and me. Yet President Clinton had made similar promises a decade and a half earlier and, because of congressional resistance, had been unsuccesful in any meaningful regulatory or legislative efforts to reduce greenhouse gas impacts.

Obama would face the same wealthy, sophisticated, and battle-hardened network that had alternately controlled and undermined the environmental policies of previous administrations. They were prepared once again to subvert efforts to combat the greenhouse gas emissions causing global warming. Within the first few weeks of his administration, President Obama called their bluff.

# PART TWO

# THE BIG DEAL

# 4

# MY NEW BOSS

"Margo Oge?"

It is 10:30 the morning of January 26, 2009. It's 27 degrees outside, and I am stuck at the White House's west entrance haggling with the guard.

"That's right. O-ghey. O-G-E." I respond. His kiosk looks heated. The bullet-proof glass is partially steamed up.

"I don't see your name."

"Maybe it was deleted during the Bush years," I suggest. The guard doesn't laugh.

After some phone calls, my ID is returned and I join the new Environmental Protection Agency administrator, Lisa Jackson, inside the White House's East Room. Lisa is from New Orleans, and I find her charming and warm. Like me, she began at the EPA many years earlier with an engineering degree. In 2002, she moved to New Jersey's Department of Environmental Protection, eventually becoming the commissioner, and later the chief of staff of Governor John Corzine. We had met for the first time a year earlier at a conference in New Jersey. Since then, Lisa has been sworn in as the EPA's first African-American administrator.

The East Room of the White House is a large, magnificent space with three-foot crystal candelabra hanging from the high ceilings. The ornate molding and massive mirrors topped with brass eagles make it feel like the kind of place where heads of state sign treaties ending bloody wars.

"Hello, Dan," I say to Dan Becker, the director of the Safe Climate Program, who was previously with the Sierra Club. There is no shortage of environmental policy experts and lawyers mingling with my colleagues from EPA today. During the previous administration many of their organizations had sued us. I didn't take it personally, of course. Just about every stakeholder in environmental issues has taken the EPA to court at some point.

Soon we all sit down, and President Obama strides up to the lectern with Lisa and the secretary of transportation, Ray LaHood. The president was inaugurated less than a week ago in front of a record crowd. He has the highest January approval rating of any post–World War II President. So it is not surprising that his long figure projects a fresh glow of enthusiasm and ambition. He smiles broadly at the audience's applause. Then he says a lot of what I've been waiting to hear from a president.

"America's dependence on oil," he says, "is one of the most serious threats that our nation has faced." He goes on to list funding dictators and terrorists, stifling innovation, and leaving the American consumer stranded as some of this dependency's negative consequences. I smile, but he is just getting started.

"These urgent dangers to our national and economic security are compounded by the long-term threat of climate change, which, if left unchecked, could result in violent conflict, terrible storms, shrinking coastlines, and irreversible catastrophe." He appears to be speaking from memory now—not glancing down at his speech. "These are the facts, and they are well known to the American people."

What is also well known to the American people is how difficult it seems to be to actually do anything about any of these problems. This president's first actions to back up his lofty words are to sit at a wooden desk set off to the right of the stage and sign two executive orders packed with Washington-speak. The first instructs the National

Highway Traffic Safety Administration (NHTSA) to finalize its Corporate Average Fuel Economy (CAFE) standards for 2011 and reevaluate the Bush proposal for new standards for the years 2012 to 2015. The second memorandum directs my team at the EPA to reconsider California's waiver request to set its own greenhouse gas standards for cars.

Both sound dully bureaucratic. Understanding why the work done in Washington matters often requires a little imagination. First, look beyond the alphabet soup of agency names. Get past words like "standards" and "waiver." Then you can see the good stuff: Americans breathing cleaner air and spending less money filling up their cars.

The president's first directive, the fuel economy standards, had proven incredibly resistant to progress for decades. Many of the same corporate interests opposed to federal action on global warming had used their considerable political clout and financial resources to stymie any rises in fuel efficiency. In fact, it took a war for us to get mile per gallon standards in the first place.

★   ★   ★

In October of 1973, the Egyptian and Syrian armies attacked Israel in a bid to regain territory lost several years earlier. The United States supported Israel in the war, provoking the Arab oil-producing nations to cut off petroleum exports. After decades of readily available and cheap gas, Americans soon realized exactly how vulnerable they were to the whims of petroleum-exporting nations.

With taps mostly shut off to cheap oil from the Middle East, Americans spent hours in line trying to fill their tanks, turning filling stations into battlegrounds. Angry customers assaulted attendants at gas stations that ran out of fuel. "Staying in line" was a mantra and a contest. People packed lunches and knitting. Several cars were totaled by a passenger train when their drivers refused to give way in a gas line that snaked across the tracks. Some states limited gas purchases to every other day.

The embargo was eventually lifted, but the issue of fuel economy suddenly had everybody's attention. In 1975, Congress instituted what are called Corporate Average Fuel Economy standards. Known as CAFE within the government, the rules required manufacturers to meet an annual miles per gallon target beginning in 1978. These standards did not apply to any particular car but were a sales-weighted mean of all the cars in a manufacturer's fleet. So, in 1978, for example, General Motors would be required to have a fleet-wide average of 18 mpg for all its new cars. These standards had a tremendous impact on the fuel efficiency of passenger vehicles in America. After decades of relatively stagnant fuel economy levels, the CAFE standards raised the efficiency of passenger vehicles nearly 10 mpg in just seven years, a more than 50 percent increase. The rules saved Americans money at the pump and moved the nation toward greater energy security.

But, as successful as the fuel economy standards were during their first seven years, the oil crisis that generated them eventually began to fade in the public's mind. In the antiregulatory atmosphere of the 1980s, with gas prices falling and pressure from the auto industry growing, America's fuel economy took a step backwards. In 1985, Ford and General Motors successfully lobbied for the standards to be loosened. From 1986, the fuel economy of cars flatlined for the next two decades. Any efforts to jump-start improvements in fuel economy for America's passenger vehicles met with fierce and bipartisan resistance.

Sitting a few rows from me is Dan Becker, the former Sierra Club director. For the past two decades he had encountered resistance from all corners in his bid to raise fuel economy. In 1989, Dan managed to get two Republican senators to sponsor a measure to bring cars up to 40 mpg over ten years. The legislation was summarily filibustered by the two Democrats from Michigan, Donald Riegel and Carl Levin, as well Don Nickles and David Boren, the Republican and Democratic senators from Oklahoma, respectively. The two states, different in geography and demographics, are united in their distaste for fuel economy rules. Michigan is the home of the Big Three car companies and thousands of union vehicle manufacturing jobs, while Oklahoma is one of the top oil producers in the nation. Though Dan got a majority of votes in the Senate, his group was unable to get the 60 votes necessary to break the filibuster and the bill died.[1]

A few years later, President Clinton explored raising the CAFE standards but was advised against the move by White House staffers. They feared the political ramifications of any regulation that was likely to be unpopular in battleground states like Michigan, Ohio, and Missouri, all of which have large numbers of autoworkers.

During the Clinton administration, Dan served on what was called the Car Talks Commission, a panel of environmentalists and auto industry representatives that was supposed to advise the president. He authored a report recommending higher fuel standards, the majority opinion of the group. But because Dan's document was not supported by the entire group, the White House angrily accused him of not producing a consensus.

Adding insult to injury, the existing fuel economy rules also got hit by Republican senator Tom DeLay. In 1975, Congress had charged the National Highway Traffic Safety Administration, an agency located inside the Department of Transportation, with setting and enforcing the fuel efficiency rules. As the agency responsible for Corporate Average Fuel Economy (CAFE) rules, NHTSA was often under assault by the auto industry and other antiregulatory interests. DeLay continued this tradition.

In 1994, taking an existing transportation bill, the senator attached a rider freezing specific parts of the National Highway Traffic Safety Administration's budget. The rider didn't actually get rid of the CAFE requirements, but it did gut NHTSA's technical resources and staff. In essence, the budget cuts meant that the agency didn't have money to even consider a marginal increase in the miles per gallon required of Americans' cars and trucks. The CAFE rules were stuck on autopilot for decades.

★   ★   ★

The second directive President Obama signs today represents a potentially huge breakthrough in the fight against global warming. To date, limiting greenhouse gas emissions has proven even more hopeless than improving fuel economy. This is striking because probably everybody in this majestic room realizes that human-created greenhouse gases are causing the earth to heat up rapidly, disrupting and exaggerating existing weather patterns. In fact, 2008, the year of his election, tied with 2001 as the eighth warmest year on record since 1880.[2] Our government's continued lack of action is locking in a series of cascading consequences for many decades down the road.

I am particularly frustrated because I know that EPA is up to the challenge. We have the skilled and motivated engineers, scientists, economists, lawyers, and policy experts to make a difference. The EPA also has experience in driving a revolution in air quality. Over the previous forty years, the agency had made American cities remarkably cleaner and our people healthier. EPA's regulations both made the air cleaner and fostered the invention of technologies like the catalytic converter, which has been adopted by the rest of the world. I want a chance to begin the same massive effort against global warming, but we had no real chance to pursue this during the Bush era. Obama's second directive may help us start a second revolution, this one aimed at greenhouse gas emissions from vehicles.

At first glance, the president's second order—reconsidering California's waiver request—may seem underwhelming. For one, it doesn't require the EPA, which is now his agency, to take direct action on climate change. Yet the order's circuitous legal logic may eventually have a similar result.

From its inception, the federal Clean Air Act law had given California the authority to set its own, more aggressive emission control standards for cars and trucks. Over the years, as many as fourteen other states, at different times, chose to follow California's standards, causing car companies to manufacture autos that met two different sets of standards. However, California was required to have the EPA's final approval before their standards could be legally enforced.

In 2004, consistent with its traditional role, California adopted aggressive regulations on greenhouse gas emissions from new cars and asked EPA for approval. In 2007, under the Bush White House, the EPA had denied the California's initial request, after delaying action for several years. Now President Obama was ordering us to reconsider the denial. If my agency approved the waiver, then the country's largest auto market would insist that cars sold there meet certain greenhouse gas emissions regulations. What's more, once California instituted these rules, other states could follow them, putting even more pressure on car companies to produce vehicles that would begin to slow global warming. Something as mundane as approving a waiver would actually be the single biggest step ever taken in the United States toward tackling climate change.

★   ★   ★

While he signs these two memoranda, Obama carries on a joking conversation with an audience member.

"You want this pen?" he asks.

The audience laughs. "Hey, Lisa, does she get this pen?"

"All right. There you go. These are nice pens"—the president is hamming it up now—"although they're a little hard to unstick."

It has been fun, maybe even therapeutic, listening to the president officially put EPA back on track. But as I pass back through White House security, I've got a larger mission on my mind.

★   ★   ★

The next morning, I push the power button on my Prius. There is no ignition sound as the lights come on, but a quiet hum as the battery powers up. I silently glide down the steep, wooded driveway outside my house in Virginia. Inside the vehicle, an electric motor augments a small 1.4-liter engine, about half the size of the engine in a Toyota Tacoma pickup. The car's iconic flat-back profile comes from incorporating a Kammback body, a design developed in the 1930s by a German aerodynamicist named Wunibald Kamm to reduce drag. The Prius is the most popular hybrid vehicle in America. Over 4.8 million Prius models are on roads around the world and account for 67 percent of the 7 million Toyota/Lexus hybrids sold as of November 2014.[3]

I'd been driving an old BMW 5 series before my Prius, which meant that my miles per gallon virtually tripled when I switched. Like other German luxury cars, my old BMW focused on performance, but it got a pathetic 16 mpg in the city.

Twenty minutes later, I turn left off the forested GW Parkway and cross the Potomac River. Memorial Bridge drops me in Washington, DC, at the west end of the National Mall. I circle the Lincoln Memorial and follow Constitution Avenue as it threads between the Washington Monument and the back of the White House. Sometimes the traffic is horrible, and I change my route.

The Environmental Protection Agency is a block-long neoclassical behemoth from 1934 with thirty-foot pillars, heroic murals, and a seven-story marble spiral staircase inside. It used to house the Post Office and is named after an ATF agent who was killed during an undercover drug sting. It is an easy walk to the White House from here. These majestic, white, columned buildings in official Washington remind me of growing up in Athens.

I ride an elevator up to the sixth floor and walk into my office. There are two windows, both of which look down into a circular courtyard. When it's nice out, employees eat lunch there. Sometimes auto companies park new cars out there for us to test drive.

On the shelves between the windows sit pictures of my two girls and my husband. There are official photos as well: President Clinton and former EPA administrator Carol Browner as they congratulate me on a Clean Car Program in 1999; President Bush giving me the President's Meritorious Service Award in 2004. Ironically, given his environmental record, this was one of the biggest awards I got from any of the five presidents I served. One of my favorites came from my

former boss in the Senate, John Chafee: "To Margo Oge with appre-
ciation for the excellent job you did. Your imaginative suggestion for
bio-degradable six-pack holders were top notch. John Chafee. May
16, 1986."

In the hallways, people are still on a high after the president's
speech. I sit down. My days are generally a parade of meetings and
always begin with an 8:30 meeting with my staff. Later that day, my
outspoken senior policy analyst, Maureen, emails me about the president's speech. "Sounded like we are back in business again."

Our business is to make sure everything with an engine moves
as cleanly as possible. The Office of Transportation and Air Quality monitors everything that moves and produces emissions: trains,
planes, lawnmowers, forklifts, jet skis, tractor trailers, farming harvest-
ers, Hummers, and the Prius I drove here today. The office has long
been a catalyst for change, forcing the development of new technologies that eventually spread around the world.

But our successes also explain why we have been fiercely and
consistently resisted by some of the largest companies on the planet,
including those in the oil and auto industries. For the past eight years,
lobbyists and representatives of these businesses had just about set up
shop in the White House. Now I'm hoping our turn has arrived.

★   ★   ★

A few days later, I meet administrator Lisa Jackson's new counsel.
She is also named Lisa, but fortunately "Heinzerling" doesn't sound
anything like "Jackson." A very sharp environmental lawyer with an
infectious smile, Lisa Heinzerling had been in DC working with
the Obama transition team since before the election. We had first
met in November, soon after Obama's victory. The outgoing Bush
administration had given Lisa a small office space on the first floor
of the EPA building. It was there that I explained to her that, if the
president wanted to reevaluate the California waiver, we already had
all the technical and legal analysis ready to go. I also mentioned the

possibility of a national greenhouse gas program that could set standards all the way out to 2025. I wasn't allowed to share any papers or detailed information then, because Bush was still in office. Now Lisa is going around to the different offices at the EPA to get a detailed briefing on our ideas.

My director, Chet France, and his technical advisor, Bill Charmley, have both flown in from the EPA labs in Ann Arbor, Michigan, for this meeting. Chet is bearded, competitive, and my right hand on technical matters. Bill is a very soft-spoken but highly competent engineer with glasses and a white goatee. Together, we lay out the plan that my team has been working on for the past two years.

We explain to Lisa that we have developed a blueprint for a single, unified national program to reduce greenhouse gases. The program we describe meets the requirements of the Clean Air Act, a 2007 energy bill called EISA, and the 2004 California Clean Car Program that is currently awaiting EPA's approval. For the first time since the Clean Air Act became law—nearly forty years ago—the United States would have one common set of emissions standards for cars for all the states. Furthermore, the National Highway Transportation and Safety Administration's CAFE standards and the greenhouse gas standards would be harmonized in a single, national program. The auto companies wouldn't love every part of the plan, but they would certainly welcome the guaranteed regulatory stability that allowed them to build vehicles that would be compliant across the whole American market.

As another part of our groundwork, some of the best automotive engineers on Chet's team—or anywhere—have analyzed existing automotive technologies. Chet knows each of the automakers' capabilities to reduce greenhouse gas emissions. Our technical and cost analysis supports taking California's new greenhouse gas regulation program to the national level with some small changes. Because lawsuits from regulated industries and the environmental groups are so common, every proposed environmental rule has to have a rock-solid legal basis. One of my favorite EPA counsels, John Hannon, and his team have already done the analysis that supports our proposal.

With the blessing of the new administration, I tell Lisa, we can go well beyond simply approving California's groundbreaking greenhouse gas law. Lisa's eyes brighten as Chet and I finish explaining our plan.

"Oh, my God!" she says. "This is the best thing I've heard from EPA since I've been here."

★   ★   ★

The EPA administrator's office is a beautiful, spacious room. It also has personal significance to Lisa Jackson, because it had once served as the Postmaster General's office, and her father had been a mailman. Several days after our discussion, Lisa Heinzerling and I are here to brief the administrator on the national greenhouse gas plan. On the wall hang some paintings that the National Gallery of Art loans out to certain offices. I see Lisa has switched out some of the works that the previous administrator had hung.

We all walk into the "bullet room," a long conference space named for the curved set of windows that look out over the office buildings lining downtown DC's 12th Street NW. This meeting is the first of a series of hurdles we have to get past before we can have any hope of turning our proposal for a national greenhouse plan into an enforceable rule.

Sitting across the conference table from the administrator, Lisa Heinzerling and I lay out our plan: one national program that satisfies all legal, technical, economic, and policy requirements. Lisa asks how I think the car companies will react.

"I think we can bring them on board," I say. "They really want a single national program." I pause. "We'll probably have to give them some flexibilities early on."

Lisa is engaged and attentive, and before too long she is committed to make it happen. This is definitely as easy as it will get.

★   ★   ★

Outside is another blustery January day, but for the first time in eight years—nearly half the time I have been running this office—I feel that we are about to come in from the cold. We had some environmental successes during the Bush years but were blocked at every turn when it came to regulating greenhouse gases. With Lisa Jackson's support, we are one step closer to finally taking action on a problem that has evaded the federal government for decades. My adrenaline begins to flow and I feel a warming optimism.

This chance for a real breakthrough has come about because of a rare alignment of mandates, laws, circumstances, and opportunities. The net result is that we now have at our disposal a collection of tools we can use toward the goal of reducing greenhouse gas emissions from automobiles. Some of these tools are already well established: One is the Clean Act Air, which has empowered the EPA since the 1970s and has dramatically improved the quality of the air Americans breathe. Another is our Ann Arbor lab, which over the past several decades has become a technical powerhouse with unparalleled engineering experience. The Ann Arbor lab has already sparked significant actions to reduce emissions from the transportation sector. But it is the more recent events that make this moment so exceptional. A recent Supreme Court decision that ordered the EPA to review the public health and welfare dangers of greenhouse gases in the atmosphere changed the legal dynamics of regulating these gases overnight. Also, fuel costs have shot up sharply over the course of the decade, leaving the near-bankrupt US auto industry saddled with an unappealing fleet of gas-guzzling SUVs and heavy trucks—and, because of the Great Recession, on federal life support. The California Clean Car Program is another tool. With our approval, California will introduce a program that will put intense pressure on automakers seeking to sell to that state's enormous market.

As unlikely as it may seem, this opportunity is also thanks in part to the Bush administration's unwillingness to take any climate-change action over the last eight years. Our office accumulated a vast amount of the engineering, policy, and legal work that goes into regulations, which we were obliged to leave on the shelf. At this moment, years of my team's work are just sitting there, at our disposal and ready to be written into a new rule.

Of course, the green light for this action comes from the president himself, but even the support of a hugely popular chief executive won't make it a cakewalk. Over the coming months, we will face longtime adversaries of greenhouse gas regulations, people who question our priorities in the midst of an economic crisis and some who won't trust our science and cost-benefit data. There will be others, even within the new administration, who are ideologically opposed to the regulations—as is almost inevitable in any room filled with Washington lawyers and academic economists. Though we have an impressive array of tools to combat the threat building up all around us, we will need all of them to move forward our historic first national greenhouse gas regulation.

# 5

# THE SECRET WEAPON

Environmental action on huge threats to Americans' health and well-being hadn't always taken so long to materialize. In the early 1970s, Senator Muskie's Clean Air Act and President Nixon's EPA had proven formidable adversaries in the fight against air pollution problems that had once seemed intractable. In comparison to today's long-stalled efforts to combat climate change, there was one big advantage back then. The pollution we were fighting was very visible.

★ ★ ★

By the middle of the twentieth century, toxic smog hung thick over every major urban center in the United States. For many, it had become a background noise for the urban experience, but it was also poisoning millions of Americans. The rancid, dirty air caused allergy attacks and asthma in children, emphysema in nonsmokers, and premature death for many others. Though these illnesses were a continuing health threat to millions, what got the headlines were the occasional apocalyptic flare-ups.

On Thanksgiving 1966, for example, a stagnant mass of warm air kept New York's pollution near street level, shrouding the city in smog.[1] The levels of sulfur dioxide, which causes difficulty breathing and has other adverse effects on the respiratory system, rose to two and half times normal. Measurements of the two other types of pollution the city monitored, carbon monoxide and smoke particles, also shot up. Aerial pictures of the city show the skyscrapers in midtown Manhattan peeking out from a dirty blanket of pollution that completely obscured shorter buildings.

The next day the stalled air sat thick around the city. The poisonous atmosphere was getting worse. The mayor ordered all eleven municipal incinerators shut down, and local utility companies were told to burn natural gas instead of fuel oil. Residents were asked to keep the thermostats no higher than 60 degrees. A city health official advised anyone with ailments ranging from emphysema to a bad cold to stay indoors. By Saturday, when a changing weather pattern finally lifted the pollution away, the smog had killed over 120 citizens. This was hardly an isolated incident. Similar events in 1953 and 1963 in New York alone were blamed for the deaths of between 300 and 450 people.

Until the Clean Air Act was passed, there were no federal limits on the rogues' gallery of gases that killed and poisoned residents. Carbon monoxide, for example, is an odorless, colorless gas, which is a product of incomplete combustion of carbon-rich materials like coal or gasoline. When humans breathe in the gas, it displaces the oxygen essential for the heart, brain, and other organs. Carbon monoxide poisoning from lack of oxygen results in headaches, nausea, comas, respiratory failure, and death. Sulfur dioxide is a colorless, sour-smelling gas produced from burning fossil fuels that, combined with nitrogen oxides, reacts in the atmosphere with water and oxygen to form acid rain.

The other pollution that New York City measured in the 1960s was the fine bits of matter that constitute soot or smoke particles. Though particulate matter is not a gas, its finest particles are actually more deadly. At 2.5 micrometers in size—a human hair is roughly 70 micrometers wide—these bits of water vapor and microscopic solids can penetrate deep into lungs and sometime even into the blood stream. They cause heart attacks, decreased lung function, and even death.

The Clean Air Act went further than New York's monitoring system by measuring an additional three pollutants and placing restrictions on them. The sweet-smelling lead that burned off from gasoline in the 1970s accumulates in the bones and affects the oxygen-carrying ability of blood. The resulting reduction in oxygen could result in damage to most of the body's internal systems and is known to permanently reduce children's mental capacity.

Another pollutant, ozone, forms a protective layer around the earth when high up in the troposphere. But ground-level ozone—what we breathe—is a major contributor to smog and causes serious respiratory illness and premature death, especially for people with lung

**Figure 1. The Clean Air Act: Forty-Plus Years of Pollution Reduction Amid Economic Growth**

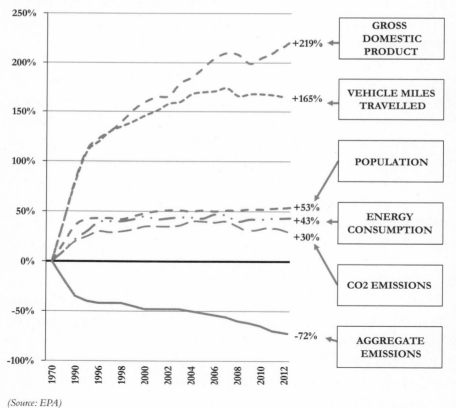

*(Source: EPA)*

disease, children, older adults, and people who are active outdoors. Ground-level ozone is not emitted directly into the air but is created when nitrogen oxides and volatile organic pollutants emitted by cars, power plants, refineries, and other sources chemically react in the presence of sunlight.

The Clean Air Act targeted each of these hazards, producing enormous health benefits. Roughly forty years after the law was passed, the aggregate national emissions of these six common pollutants had dropped by an average of 72 percent, leading to significant improvements in health while vehicle miles traveled increased 16 percent and the economy, as reflected by the GDP, grew 219 percent (Figure 1). In 2010 alone, the CAA is responsible for saving over 160,000 lives and preventing millions of cases of respiratory problems. Also, an independent study showed that reductions of fine particle pollution alone in American cities between 1980 and 2000 led to improvements in average life expectancy at birth of approximately seven months.[2]

That the law improved Americans' health is not surprising, but the measure also created huge economic benefits. For example, employees with better health have fewer medical expenses and lower rates of absenteeism. Although the EPA's rules were designed to create maximum health benefits for Americans while keeping companies competitive and profitable, industries that had been previously unregulated for environmental concerns still fought tooth and nail against every proposed regulation. One the first big battles EPA faced was with the automotive industry over cleaning up the filthy emissions from American cars and trucks.

★   ★   ★

In 1970, the same year that the Clean Air Act was passed, a twenty-seven-year-old mechanical engineer named Mike Walsh began working on New York City's new environmental initiative. With help from an EPA grant, the city had set up a lab to study air pollution and possible ways to mitigate it. Mike joined a project testing a new device that could be attached to a car's exhaust system to limit the various toxic fumes spilling out of the tailpipe. With auto emissions yellowing the sky and making breathing difficult and hazardous, the catalytic converter was getting attention as a relatively easy and cost-effective way to clean up cars.[3]

Mike's team placed catalytic converters on municipal vehicles like gas-powered garbage trucks and police cars. The devices did their job, cleaning up  emissions substantially. The team figured that if the devices would work on police cars alternately idling and tearing down pot-holed streets in both snow and the August heat, they'd work just about anywhere. These were the only non-industry-conducted tests of catalytic converters, and Mike shared the data with the EPA as the agency geared up for a showdown.

On the morning of March 12, 1973, Ernest S. Starkman, a vice president at General Motors, sat behind a felt-covered table. Opposite him at an identical table was the first administrator of the Environmental Protection Agency, William Ruckelshaus, his lips formed into a flat, determined line under black, horn-rimmed glasses. Each man was flanked by his colleagues and surrounded by reporters from newspapers and television networks. The venue—the largest open space in the Department of Agriculture's Washington, DC, headquarters— looked like a high-school auditorium.[4]

The TV cameras rolled as Starkman made his case that compelling General Motors to clean up its tailpipe emissions in the next two years would be a dangerous mistake. The EPA mandate requiring automakers to reduce car emissions forming smog and carbon dioxide by 90 percent on their 1975 model year cars presented "an unreasonable risk in business catastrophe."

"It is conceivable that the complete stoppage of the entire GM production could occur," said Starkman, "with the obvious tremendous loss to the company, shareholders, employees, suppliers and communities."

Ruckelshaus stared back, apparently unimpressed. "Why can't you make it? If some companies can make it, why can't all of them?"

Ten days earlier, a reengineered Mazda had already passed the stricter 1975 emissions test in the EPA's Ann Arbor lab.[5]

The next day, Ford and Chrysler executives took seats behind the same table to weigh in against the mandate as well: the current technology didn't work, it caught on fire, it reduced engine performance substantially. Outside the auditorium, the nation's capital sloughed brown in smog. Ruckelshaus held firm.

Two years later, the largest of what were then the Big Four automakers (American Motors Corporation was the fourth) took out newspaper ads with a headline that happily boasted, "General Motors Believes It Has an Answer to Automotive Air Pollution Problem . . . and the Catalytic Converter Has Enabled GM Engineers to Improve Performance and to Increase Miles Per Gallon."[6] The copy went on to list the virtues of the new technology: the technology is safe, it is cleaner, it doesn't reduce performance, and it improves efficiency. In a paid ad, the automaker refuted virtually every argument against the technology it had made in the 1973 hearings.

The introduction of the catalytic converter was a highly successful example of the kind of revolutionary changes the EPA regulations in the 1970s were capable of fostering. The agency set tough standards to clean up the air: a 90 percent reduction in hydrocarbons and carbon monoxide by 1975, with a similar reduction for nitrous oxide by 1976. But what it didn't do was just as important.

The EPA didn't tell automakers what technology they had to use to make the improvements. It didn't pick winners and losers. Instead, the mandate created a huge market for whatever new technology could get the job done. Private industry would have to figure out the rest. In the case of the catalytic converter, it came down to a frantic battle between Corning Glass and Engelhard, a New Jersey based chemical and metal smelting business, to grab the millions of dollars in contracts from auto companies.

Once the technology had been developed, another problem cropped up. At the time, all gas increased its octane by adding lead. But lead also gummed up the chemical reactions that allowed catalytic converters to clean auto exhaust. After only about 10,000–15,000 miles, the catalytic converters would be nearly useless. For the new air-cleaning technology to work properly, the United States would have to get rid of leaded gas. So, over the next couple decades, leaded gas disappeared from America's filling stations.

The catalytic converter removed an unprecedented amount of chemicals from the air Americans breathe. More surprising, perhaps, is that it achieved these reductions in the United States long before Europe and the rest of the world. In 1990, when the vast majority of American cars had catalytic converters installed, only about 12 percent of European cars did. Today it is hard to find any cars in the world without this component. Ironically, the emissions standards that American automakers resisted so vehemently in the early 1970s made American businesses world leaders in clean emissions technology.

The catalytic converter's success exemplified what the EPA was supposed to do. Clean up America. Save and improve the quality of lives. Create new jobs and innovative industries. The agency, started by a Republican president and supported by his successors from both parties, may have met resistance from certain industries in 1970s, but its work was not overly politicized during its first decade. Then, in late 1980, Ronald Reagan was elected president and progress came to a grinding halt.

★   ★   ★

In 1977, Mike Walsh, the engineer who had worked on the early catalytic converter tests in New York City, took over the EPA Office of Mobile Sources, holding the same job I would nearly two decades later. For three years, he thought he had the best job in the world. Then, on a Sunday in early 1981, Mike found himself standing in the EPA office of his new boss, Anne Gorsuch, explaining the new technology behind the three-way catalytic converter.[7]

After about ten minutes he paused. "Madame Administrator, do you have any questions?"

"No," said Gorsuch. "Just go ahead, Mike."

Mike continued talking for nearly an hour about every new issue facing his EPA office, which was in charge of emissions from cars, trucks, and other vehicles. Auto manufacturers had come to the new administration with a laundry list of requests for relaxing environmental regulations. Methodically, Mike went through and explained his office's position on each. No, carbon monoxide regulations should not be relaxed. No, nitrogen oxide standards should not be relaxed. No, inspection programs should not be made voluntary. He was calm, thorough, and speaking from experience.

When Mike finished, Gorsuch turned to the temporary administrator. "Who's next, Walt?" Mike sat and waited for all the other EPA senior leaders to give their speeches to the politely unreceptive Gorsuch.

The next Sunday, the administrator called all the heads of the various EPA offices back to her office. "OK, Mike. Thank you for your briefing last week. Here's what we're going to do. We're going to recommend that standards for carbon monoxide and nitrogen oxide be relaxed. We're going to roll back the inspection programs."

"But, Madam Administrator," said Mike. "If we do this, do you understand—?"

"I understand Mike, but this is what we're going to do."

Mike resigned and left EPA a few months later. In 1982, the administrator—now married and named Anne Burford—became the first cabinet-level head of an agency ever cited for contempt of Congress after she refused to provide documents in an investigation. She resigned soon thereafter, in disgrace, leaving behind an agency described by an EPA official as "a shambles." To regain credibility, President Reagan asked the first administrator, William Ruckleshaus, to return. Nonetheless, Burford remained proud of cutting the EPA's budget by 22 percent in just twenty-two months.

Despite the heavily politicized atmosphere at the EPA beginning in the early 1980s, the Office of Mobile Sources made good use of the Clean Air Act. The office was successful in continuing to reduce the lead content from gasoline, reducing pollution from cars, and improving the environmental performance of gasoline fuels. Amendments to the Clean Air Act in 1977 and 1990 broadened its authorities to regulate other sources beyond cars, including commercial trucks and buses. Yet, while the Clean Air Act, our founding document, was our most important tool and gave us the legal standing to pursue health benefits, another—technical—marvel was indispensable in the effort to combat air pollution. Our secret weapon was a concentration of expertise in engineering, science, and policy sited in a Midwestern city. No one else in EPA had anything like it.

★　★　★

Since the early 1970s, the EPA had run a testing station out of a warehouse in Ann Arbor, Michigan. Less than a week after moving to the top spot at the Office of Mobile Sources in 1994 (it was renamed the Office of Transportation and Air Quality—OTAQ—in 1996), I took my first trip out to my National Vehicle and Fuel Emissions Laboratory (NVFEL) located on the northwestern outskirts of Ann Arbor. After my flight, I drove about thirty minutes west from Detroit's airport.

This EPA lab is located in Ann Arbor for two reasons. First, the facility is about forty-five minutes away from the various headquarters of the Big Three automakers, Ford, GM, and Chrysler. The other advantage is that Ann Arbor is the hometown of the University of Michigan. In the early 1970s, when the lab opened, the EPA had no computers to run tests but, through the university, lab technicians were able to get access to what was then the most advanced computing technology. Today, of course, the lab has its own computers, but its location allows it to recruit top engineering graduates from Michigan as well as hire employees from Ford, GM, Chrysler, and other vehicle manufacturers.

I parked outside the hangar-like building in a space where suburbs, malls, and industrial parks all converged and walked through a parking lot full of SUVs and big pickups. Inside the lab's front doors was a huge open area with a shiny poured cement floor and exposed duct-work that crisscrossed a thirty-foot-high ceiling. A casually dressed crowd of men and women gathered around me in a semicircle.

"Hello," I said, looking around at my new team. One techni-cian had on a cowboy hat and boots. I gave it to them straight. "I don't know much about cars, but I am a fast learner and deter-mined to work with you to write the most protective public health regulations."

I had heard that the staff was very concerned about my tendency to move from job to job within EPA. The previous office director, Dick Wilson, had been in charge of Mobile Sources since Mike Walsh left in the wake of the Reagan-era attack on environmental regu-lation. Dick's tenure stretched over fifteen years, or half the office's existence. People there had grown up with him, and, in the face of intense political opposition, had tried to continue the progress. "The longest I have stayed in any job is three years. When I feel the job is done, I move to a more challenging job, so I don't know how long I am going to be here, but I am very excited. Together we are going to write environmental history."

While I had their attention, I laid out my management style. "I have heard that most of the meetings are done in small groups and that staff are not allowed to participate. I want to have open meetings, and I want people to participate. At end of the day, our challenge is to protect public health, and we have to do it together."

To some I might have seemed an unlikely boss at the lab. I wasn't an air or auto expert. I was a woman with a Mediterranean cadence to her voice taking over a lab in Michigan full of guys who love cars. But at least I was an engineer—and a good manager.

★  ★  ★

The lab that has played such a critical role in cleaning up the nation's air is divided into twenty-three different testing sites for engines, vehi-cles, fuels, and batteries. The vehicle testing team takes "secret cars," generally the next year's models of cars and trucks, and drives them on top of huge rolling cylinders called dynamometers. These devices are sort of like fancy vehicle-sized exercise bikes—no matter how fast the wheels spin, you never go anywhere. Their purpose is to simulate real road conditions like speed, acceleration, weather, incline, and air conditioning usage. The testing confirms the fuel economy and emissions values for new vehicles before they enter the marketplace.

At the engine testing sites, technicians evaluate the emissions of everything from tractor-trailer engines to lawn mowers and chain saws, sucking the exhaust into analysis chambers. The fuels testing team evaluates the quality of gasoline, diesel, and alternative fuels while the battery testing center assesses the driving range of electric vehicle batteries.

The lab is also where the EPA city/highway fuel economy sticker that appears on the window of every new car at a dealership originates. In 1975, when Congress gave the job of setting and enforcing the first-ever CAFE (Corporate Average Fuel Economy) standards to the National Highway Safety and Transportation Administration, there was a twist. NHTSA, pronounced "Nit-sa," is best known for running cars into walls and then seeing if the crash test dummies survived. With the CAFE rules, NHTSA wasn't in charge of testing how much fuel cars and trucks were actually burning: it didn't get to test its own standards—to smash cars into the wall for fuel economy, so to speak. Because EPA was already monitoring vehicle exhaust for pollutants, our lab got this job too.

From a technical and cost effectiveness standpoint, the decision made a lot of sense. Gasoline is carbon-based, so EPA lab technicians simply recorded the amount of carbon coming out of tailpipes they were already testing for other pollutants. Once they knew how much carbon a car was burning under various conditions, it was easy to calculate a basic miles per gallon. Then we provided the test data to NHTSA for enforcement.

But our testing procedures weren't perfect, and, over the course of my years overseeing the Office of Transportation and Air Quality, there was a widening reality gap between the mileage displayed on EPA stickers at dealerships and the actual fuel economy of those cars once they got on the road. Because gasoline prices remained low and fuel economy was not a major concern at the time, this discrepancy never got any major publicity.

For me, this slippage came home in 2003 when I bought my second generation Toyota Prius. I was excited to drive a vehicle that the EPA window sticker said would get 60 mpg in the city and 51 on the highway. But during my first winter driving the Prius, it started to bother me that the mileage was so far below what the sticker promised. Instead of getting close to 60 mpg, I was getting less than 38 mpg. In cars that promised such high efficiency, this difference—22 mpg—was too big to ignore. I took the car to my dealer and explained the problem. After couple of days they called me back: "There's nothing wrong with your car."

I called up my engineers. "What's going on? Our sticker is off by over 35 percent."

During my next trip out to Ann Arbor, I convened my engineering team to begin a review of the testing procedures. The fuel economy test simulated a 7.5-mile urban drive at an average speed of about 20 mph as well as a 10.2-mile highway drive averaging about 48 mph. The tests hadn't changed in decades—but American driving habits had. More people live in the suburbs, there's more traffic, cars drive at higher speeds, and people drive more aggressively. Just about every car comes with air conditioning too, which drastically reduces fuel efficiency. As a result of all these technical and social changes, the EPA fuel economy values were divorced from actual fuel efficiency. Reputedly high-mileage cars like the Prius just served to underscore that divergence. During the next several years, my office better aligned the mileage estimates with current driving habits and technologies and twice updated new cars' window stickers. We developed the Green Vehicles Guide (www.epa.gov/greenvehicles/) to help consumers find vehicles that are more efficient and less polluting. In collaboration with Department of Energy (DOE), we also launched www.fueleconomy.gov, where the EPA fuel economy rating of every car sold in the United States is listed. The 2004 Prius that was originally rated at 60 mpg in the city and 51 mpg on the highway is now rated at 48 mpg city and 45 mpg highway, much more in line with drivers' real-world behavior.

But we could not fix the other half of the problem—the testing procedures used to set the CAFE standards for regulating the fuel efficiency of the car companies' fleets. These standards are administered by NHTSA, and the test procedures can only be changed by Congress. When the EPA lab began to more accurately gauge fuel economy, NHTSA was not allowed to bring CAFE values in line without Congressional approval. Since Congress has so far refused to do so, CAFE values continue to be based on the antiquated 1970s test procedures that, among other things, do not account for highway driving at over 55 mph, the nationwide speed limit at the time, or the use of air conditioners. As a result, the CAFE values and EPA fuel economy values reflected in the window sticker continue to be significantly different, with the CAFE mpg values typically about 20 percent higher than the EPA ratings, needlessly confusing consumers who are interested in fuel economy.

★   ★   ★

The Ann Arbor lab wasn't just a testing facility, though. The scientists and engineers at the lab held over 50 percent of all the patents awarded to the twenty-nine EPA R&D laboratories across the country. It earned royalties on innovative technologies like hydraulic hybrids, which it developed in conjunction with private industry and now licensed.

In 1990, we got a new tool that extended the agency's—and lab's—authority: Congress handed us a freshly amended, versatile, and powerful Clean Air Act. The law gave my office broader latitude in reducing emissions not only from cars and trucks but also from non-road vehicles like locomotives, marine engines, and agricultural equipment burning diesel fuels as well as gasoline.

In the lab's offices there were also lawyers, economists, and policy experts who actually wrote the regulations that were responsible for reducing air pollution and improving public health and the environment. They knew how to work with their colleagues, the scientists and engineers, to develop technology-forcing standards that made America the world leader on automotive emissions controls. For example, the car companies' innovations in areas like onboard diagnostic computer systems were largely the result of these standards. More than that, with America's leadership, such innovations were adopted in other parts of the world. Improving air-to-fuel injection made vehicles more efficient and cleaner not only in the United States but around the globe. As we pursued new regulations based on the updated act, the lab team gave me complete confidence that we could once again develop world-leading standards.

At this juncture, my colleagues and I know that this history of world-changing innovation effectively doubles down on the bet we are placing on the national program for regulating greenhouse gas emissions from cars. Just as with catalytic converters in the 1970s, we are aiming to set a new bar for admission to the world's most competitive auto markets. If the United States demands innovative, cost-effective greenhouse-gas-reducing technologies, automakers around the globe will listen.

As I begin to think about the strategy to bring all the stakeholders in the Obama administration on board with the national program, I know how hard it is going to be, but I also knew that we stand on a firm technical footing to move forward.

# 6

# YES, WE CAN

At 5:16 p.m. on Saturday, January 30, 2009, I email Lisa Jackson's senior climate policy counsel, Lisa Heinzerling. "Going to NHTSA will not work. Can we ask them to come to EPA?"

Any national program to regulate greenhouse gases will have to include both California and the National Highway and Safety Administration. Carbon dioxide is the major greenhouse gas emitted from cars. In general, higher fuel economy standards result in lower emissions of carbon dioxide. Because NHTSA sets the nationwide CAFE fuel economy rules, they are a critical piece of efforts to move forward our greenhouse gas plan.

Unfortunately, our last collaboration with NHTSA broke off when the deputy administrator, Ron Medford—and everyone else at his agency—stopped responding to my emails one day during Bush years. In 2007, President Bush had issued an executive order that NHTSA and EPA work together on a single regulation for both greenhouse gas and fuel economy. Then the administration abruptly changed course, telling NHTSA to end all collaboration with us immediately. That unexplained rupture had left some scars between the two agencies. There are plenty of good people at NHTSA, but our awkward working relationship makes that agency a hurdle for our greenhouse gas plan. So far I haven't been able to even get a meeting with them.

Lisa gets back in a couple minutes. "Yes, we can"—unconsciously echoing the mantra of the Obama campaign—"does one o'clock still work for you?"

"Yes. (Yes We Can.) Lisa, I have heard from my people, and their sentiment reflects my views: you are such a breath of fresh air at EPA. We are blessed to have you as part of the EPA team." Campaign slogans aside, the sense of excitement and the potential for progress around my office are in stark contrast to the frustration with the intransigence of the Bush years.

The previous administration had been cold toward most of EPA's environmental initiatives, having adopted as its rigid orthodoxy the contention that there was too much uncertainty to take meaningful action on global warming. In that atmosphere, many environmentalists and other nongovernmental organizations had all but abandoned their push for federal regulation or legislation on climate issues. There wasn't much promise of action in Congress either. But there are still ways to succeed during even the most hostile presidencies. Fed up with the federal government, many environmentalists and a number of other concerned groups looked to the states. In the national leadership vacuum, places like California and Massachusetts had begun to take the lead on climate change early in the Bush years.

In fact, a Supreme Court victory won by a coalition of states, environmental, and other nongovernmental groups in a landmark case should have forced President Bush to take action. Predictably, the Bush administration did all it could to delay that action and, eventually, to sabotage EPA's efforts to respond to the court. Today, that court decision was still an important tool that provided legal leverage for my team to sell our greenhouse gas proposal to a more environmentally friendly administration.

★　★　★

The legal milestone that came to be known as *Massachusetts vs. Environmental Protection Agency* had a very unlikely father—House Republican

Tom DeLay, a steadfast opponent of environmental regulation. In 1995 the congressman from Texas had called the EPA "the Gestapo of government." Three years later, on March 11, 1998, he looked down from his raised lectern at President Clinton's EPA administrator, Carol Browner. Ever since she had taken over the agency, in January of 1991, Carol been a great advocate for my team's efforts to clean up cars, trucks, and fuels. She was also known for protecting the agency from Republican attacks with steely resolve. Now she was facing yet another one.

Carol was seated behind a desk in the middle of the Congressional hearing room during the second day of the EPA's appropriations hearings. These often-torturous events are intended to provide the heads of agencies with an opportunity to justify the money they are asking for in the following year's budget. But it also provides our congressional representatives a chance to ask tough questions about almost anything.

Representative DeLay was a persistent inquisitor. He had spent much of his question time during the first day trying to badger Carol into saying that the Clinton administration was attempting to implement the Kyoto Protocol, an international treaty, without Senate ratification. Carol had proven too tough and smart for that tactic, so the congressman came back from a slightly different angle the next day: "Do you think that the Clean Air Act allows you to regulate the emissions of carbon dioxide?"

DeLay's question was no doubt based on a recently leaked EPA memo. At around the same time as the hearing, the Department of Energy was working on legislation to restructure the electricity sector. David Doniger, a former lawyer for the National Resources Defense Council, was working as a counsel for Bob Perciasepe, the EPA assistant administrator of air and radiation. My boss at the time, Bob was tireless, humble, respectful, and a strong leader. Bob and David reasoned that if the energy sector was going to be restructured anyway, this was a good chance to ensure that EPA had the unquestioned authority to regulate public utilities for the four key pollutants—sulfur dioxide, nitrogen oxide, mercury and carbon dioxide—as a set. After discussions with EPA administrator Carol Browner, David drafted a memo. Somehow, the information had ended up in DeLay's hand. It clearly had grabbed his attention.[1]

Browner leaned slightly toward the microphone pointed at her. "I think that we are granted broad authority under the Clean Air Act to—" DeLay cut her off, demanding that she produce a legal opinion.

One month later, the EPA's general counsel, Jon Cannon, did just that. His memo affirmed that the EPA had already asserted its authority to regulate pollutants including particulate matter sulfur dioxide, mercury, and nitrogen oxides, a precursor to ozone, and the agency was well within its rights to regulate carbon dioxide, even if it had not yet exercised that power.

It was a fairly innocent five pages. The blustering conservative congressman—a man whose legislative career had earned zero rankings by both the Sierra Club and the League of Conservation Voters—could not possibly have imagined the chain reaction of events that the Cannon memo would set off.

In October 1999, a group of environmental and public health organizations led by the International Center for Technology Assessment petitioned the EPA to regulate climate change-producing emissions like carbon dioxide and other greenhouse gases from new motor vehicles. In the face of the federal government's inaction on global warming, the group was seeking to force the EPA, via the legal system, to regulate greenhouse gases under the 1990 Clean Air Act. In their petition, they cited the legal opinion demanded by Representative DeLay and provided by Jon Cannon, arguing that the Clean Air Act gave the EPA the ability to take action to mitigate the effects of global warming. In the final days of the Clinton presidency, the EPA put out the petition for public comments. By that time, George W. Bush was president-elect and the petition went unanswered for another several years.

★　★　★

On September 8, 2003, the Bush administration formally responded to the petition, denying it. This, too, was a political gambit. The administration was well aware that the coalition would sue them for denying the request. But they thought they could win, especially in the relatively conservative US Court of Appeals for Washington, DC.

The EPA's response to the petition claimed that it lacked the authority to regulate carbon dioxide and other greenhouse gases for climate change purposes. It went so far as to declare that *even if* the EPA did have the authority, it would refuse to set greenhouse gas emission standards for numerous policy reasons. The agency stated that regulating motor vehicle GHG emissions would not be an effective strategy for addressing the global problem. It also argued that mandatory efforts will not be consistent with President Bush's policies for addressing climate change, which focused on voluntary reductions in greenhouse gases. My team and I could only watch as the EPA passed up an historic opportunity to initiate federal action on climate change.

Jeff Holmsted, the assistant administrator of the Air and Radiation Office and my boss at the time, assigned lawyers from the office of EPA's general counsel to write the response. I told Jeff that none of my staff wanted their names listed as the party responsible for responding to public comments. It was a small way of signaling how profoundly they disagreed with the new agency position. Jeff asked one of the general counsel lawyers to be the face for public comments instead.

The petitioning groups were hardly surprised to have their request rejected by the EPA. In fact, the rejection was a necessary first step to filing a lawsuit. The legal action that was the next step had a much larger base of support. On October 18, 2003, fourteen environmental groups, twelve states, three cities, and the American Samoa territory filed a petition with the US Court of Appeals in DC, asking the court to review EPA's decision.  Party to the suit against the EPA, besides the territory of American Samoa, was the entire West Coast, all of New England, Illinois, New Mexico, both New York state and city, Baltimore, and the EPA's hometown of Washington, DC. The environmental organizations joining the suit included the International Center for Technology Assessment, the Natural Resources Defense Council, the Union of Concerned Scientists, the Center for Biological Diversity, Greenpeace, the Sierra Club, Friends of the Earth, the Environmental Defense Fund, and a number of other prominent scientific and environmental groups.

On the other side, the EPA position had support from ten states that formed a wide swath across the middle of the country, including Ohio, Texas, the Dakotas, and Idaho. A number of industry groups, including the Alliance of Automobile Manufacturers, National Automobile Dealers Association, and Truck Manufacturers Association also took up the EPA's cause.

As with any large group, the coalition of petitioners argued over how to make their strongest case. Massachusetts had wrapped up the lead role in the lawsuit, so the specific grievance rested on the state's territorial losses along its 900 miles of coastline due to rising seas from global warming.[2]

With several dozen lawyers involved, many of them passionately committed to their cause for decades, disagreement was often the norm. Representatives from groups like the Sierra Club, Natural Resources Defense Council, and Environmental Defense Fund argued with states' attorneys that didn't have the same background in environmental law. After a long series of contentious conference calls and meetings, they finally agreed to split up the role of writing the brief.

But all this verbal wrangling came to nothing. On July 15, 2005, the US Court of Appeals in Washington, DC, ruled in favor of the EPA in a 2–1 decision. Now the group faced a high-stakes decision.

This sequence of petition and lawsuit was like a game of poker. The coalition's initial petition tried to leverage the memo on carbon dioxide that Tom DeLay had demanded. The subsequent denial of the petition by the EPA set the stage for a lawsuit that both sides anticipated. But once the Bush administration's EPA won that case, there was a new precedent limiting the EPA's ability to regulate carbon

dioxide emissions. The coalition could try to get the case heard by the
Supreme Court, but that move would raise the stakes considerably. A
loss at the Supreme Court would severely limit implementation of
greenhouse gas regulations through the Clean Air Act. There would
be virtually no existing tools left to promote national action on climate change. Any further action would have to come in the form of
legislation from a hopeless Congress. The only way to move forward
would be to pass a new law, an unattractive option for environmentalists given the army of automobile, oil, and coal lobbyists that would
kill any legislation strong enough to have any substantial effect on
their industries.

On the other hand, a victory for the coalition would be tremendous. Again, the environmentalists and states argued about whether
to move forward. Then they pushed all their chips to the center of
the table. They were taking their greenhouse gas case to the Supreme
Court.

★   ★   ★

On June 26, 2006, the Supreme Court agreed to hear the case. Its official name was *Massachusetts vs. Environmental Protection Agency*, but everyone called it *Mass vs. EPA* for obvious reasons. Because his state had
the lead role, James Milkey from the Massachusetts attorney general's
office had the responsibility for drafting the legal brief. He decided to
ask Lisa Heinzerling, then a Georgetown University law professor, to
be the actual author of the brief.

This was a huge honor and responsibility for Lisa. From the beginning, she figured that the case was just as much about the politically
volatile nature of greenhouse gases and climate change as making
strictly legal arguments. She quickly ruled out using international
laws regarding the environment or the politicized science of climate
change as the basis for her case.[3]

To further uncouple her case from the battle over global warming, she handled the brief's wording like Ming dynasty–era porcelain.

Strategically, the petition never referred to "global warming" but to "climate change." The petitioners didn't use the term "greenhouse gases," instead substituting "air pollutants associated with climate change." There was no reference to the "Bush administration" or the "Bush EPA," but merely to the EPA. The group also hoped that the amicus briefs from diverse groups including religious organizations, climate scientists, the skiing industry, state and local governments, and some progressive energy companies might make their case seem more than a blue states versus red states argument.

Similarly, Lisa decided not to play up the case as deciding "the great debate of the day." She asked the court to simply compel the EPA to go through the normal legal procedures for determining whether or not to regulate greenhouse gas from vehicles. In her argument, the rule was clear: The Clean Air Act defines "air pollutants" as "any air pollution agent or combination of such agents, including any physical, chemical, biological, radioactive . . . substance or matter which is emitted into or otherwise enters the ambient air." Since all greenhouse gases that petitioners had asked the EPA to regulate were in fact physical or chemical substances emitted into the ambient air, the question of whether they could be added to that list seemed to be straightforward, and the answer an unequivocal "Yes."

But any flawed assumption in the legal argument, no matter how basic, could be the downfall of the case. Lisa spent days writing on this issue. One night her ten-year-old daughter asked her what she had been doing all day.

"Well, I've been explaining the meaning of the word 'including.'"

"How come," she asked, "these people who are, like, sixty or seventy years old don't know what 'including' means?"

Good question, Lisa thought, but the EPA didn't seem to know either.

Working on the other side of the case was John Hannon, an animated and creative EPA lawyer who always showed up for meetings carrying the same worn-out copy of the Clean Air Act. Now, as a member of the EPA's General Counsel Office, John was charged with preparing a defense that severely undercut the scope of that treasured document.

Of course, trial lawyers everywhere are frequently asked to effectively represent clients whom they don't agree with, sympathize with, or even believe. One difference was that many of the lawyers at the EPA had joined the agency not just for a paycheck but because of their personal values. They wanted to spend their time working on the side of science and the law and, as the agency's name suggests, protecting the environment.[4] For the EPA lawyers, the psychological strain of working long hours on a legal argument to limit their agency's authority to protect the environment was intense. Their frustration sometimes spilled over into shouting matches. Additionally, John found himself facing off against his former colleagues. He and David Donigar, for example, had worked side by side during the Clinton administration. Now they were on opposite sides in an environmental civil war.

★   ★   ★

Three weeks before the court date, the lawyer who would argue the petitioners' case before the court, James Milkey, disappeared. The rest of the coalition eventually found out he was doing individual practice sessions with a Supreme Court expert and lawyer from Georgetown University. After years of preparations and contention, thirty organizations and states went into the first day of arguments just after Thanksgiving of 2006.

Moments into his argument, Milkey was questioned by Justice Antonin Scalia.

"When? I mean when is the predicted cataclysm?" Justice Scalia demanded and then proceeded to interrupt Milkey's answers with more questions. Chief Justice John Roberts and Justice Samuel Alito followed suit. Lisa's effort to decouple the case from the politics of climate change was not working out smoothly so far. Neither did the intense questioning bode well for Massachusetts and the environmentalists.

Despite the hostile court, Milkey handled himself well, producing convincing answers for each question. Then Justice Anthony Kennedy asked him a question about what case set a relevant precedent for states' rights. Milkey drew a complete blank and paused.

"I would have thought that your answer would be *Georgia v. Tennessee Copper*," suggested Justice Kennedy. The case he referenced involved a copper smelting plant in Tennessee that was sending noxious fumes over the border into Georgia. The court had found that the smelter was demonstrably causing damage to Georgia's land—no vegetation was alive downwind of the copper plant—and issued an injunction on the plant, eventually shutting it down. This case gave standing to the petitioners to claim harm. Justice Kennedy's suggestion was a gift. Observing the oral argument, two National Resource Defense Council lawyers clenched their fists. They had the identical thought: This is the case that will bring Justice Kennedy over to our side![5]

The Supreme Court doesn't produce rulings the day or week or even month that arguments are heard. It wasn't until April 2, 2007, that I learned anything more. My BlackBerry rang during a meeting at EPA headquarters. John Hannon's number showed on the screen.

"Why is he calling?" John worked just upstairs and popped in when he wanted to talk.

"We won the case!" he shouted into the phone. "I'm coming down to fill you in."

Technically, we had lost. The Supreme Court had decided in favor of Massachusetts. But in the black-is-white world of Bush-era environmental policy, "we won" by losing the case. The court's decision would now force EPA to take action on greenhouse gases under the Clean Air Act. Seven years into a difficult era for environmental regulations, most of us celebrated our loss. I had tears in my eyes as John gave me a big hug. Before too long we were offering champagne toasts to our defeat.

I didn't mind losing to win—the politically warped logic is just part of doing business in Washington. Now, two years after we cheered our agency's loss, what really matters is that we can do our job. Many

obstacles remain. Our upcoming meeting with NHTSA, for example, will be the first test of interagency cooperation on such rulemaking in the new administration. But the *Mass vs. EPA* handed us
an important legal precedent to move forward with national law on
greenhouse gas—and beyond the unrelenting resistance of the Bush
administration. As a bonus, we had also added Lisa Heinzerling, the
very thoughtful lawyer who had helped make history, to our EPA
team. And, for better or worse, large parts of the domestic auto industry were on the verge of a historic collapse.

# 7

# FUEL ECONOMY'S REVENGE

It is Monday, February 2, 2009, and the 67th Annual Washington Auto Show is under way. The auto show always fills the Washington Convention Center with the latest car models, prototypes, salespeople, and Washington Redskins players signing autographs. This year the automakers have more than thirty green models on display, from plug-in cars to solar-paneled cars. GM is displaying its prototype for the Chevy Volt and touting its fuel savings. I walk the convention hall wondering about how many of these concepts will actually make it into production and help us to reduce greenhouse gas and improve fuel economy.

"We're hopeful it'll do what auto shows are supposed to do, which is excite the public's imagination and stimulate consumer confidence to bring buyers back to the market," says Gerard Murphy, president of the Washington Area New Automobile Dealers Association, referring to the continuing slide in auto sales.[1]

Murphy's concern is warranted; the national economy is tattered. Real estate prices have been in freefall over the past two years. The Dow Jones average halved its value in the past year, eviscerating millions of Americans' retirement funds. Financial giants like Lehman Brothers and Bear Stearns have disappeared; the government is propping up Citigroup.

American auto companies are at least as weakened as those in the financial sector. The previous November, the CEOs of the Big Three auto manufacturers, GM's Richard Wagoner, Ford's Alan Mullaly, and Chrysler's Robert Nardelli, flew in their corporate jets from Detroit to Washington, DC, to beg Congress for a $25 billion aid package to keep their money-hemorrhaging companies out of bankruptcy. Their visit was unsuccessful.

Congress refused to grant their loan request but asked them to draft a new action plan to keep their companies, and thus American auto manufacturing, solvent. Instead of financial support, congressmen from both parties offered angry ridicule. "There's a delicious irony," said New York congressman Gary Ackerman to the CEOs, "in seeing private luxury jets flying into Washington, DC, and people coming off of them with tin cups in their hands. It's almost like seeing a guy show up at the soup kitchen in a high hat and tuxedo. . . . I mean, couldn't you all have downgraded to first class or jet-pooled or something to get here?"

By the end of 2008, the government had forced Citigroup, Bank of America, JP Morgan, Goldman Sachs, and other reeling financial giants to accept tens of billions of dollars in loans. But the auto manufacturers had to make due with an emergency lifeline of cash that President Bush threw them soon before leaving office. Little else had improved for the companies since then. Their sales were continuing to crater. No one was buying new cars. Just a week ago Italian carmaker Fiat agreed to buy Chrysler's best parts at a deep discount.

The debate over whether to allow the carmakers to fail—as investment services company Lehman Brothers had the past September—or to save American manufacturing jobs is ongoing. The ultimate fate of companies that for decades were the pride of the United States is now a huge question mark. But while their financial decline was precipitous, the automakers' underlying problems had deep roots. In a sense, their current fate began not with a blunder but a huge success.

★  ★  ★

In the 1970s, Ford employees Lee Iacocca and his partner, Hal Sper-lich, conceived of a vehicle that featured the storage capacity of a cargo van sitting on top of a midsize car chassis. The idea was rejected by Ford management. A decade later, after both men had moved to Chrysler, the concept was reborn and released to the public in 1984 as the Dodge Caravan. The vehicle's success was tremendous. A year later, Chrysler was having trouble keeping up with demand, competitors were rushing their own "minivans" to market, television ads touted the vehicle as a "magic wagon," and the traditional sta-tion wagon was on its way to obsolescence. The introduction of the first modern minivan had revolutionized the family car market, but it had an even bigger impact on American driving habits and fuel efficiency.

The Caravan's less heralded breakthrough was ingeniously exploit-ing a distinction made between cars and trucks in Congress's CAFE fuel efficiency requirements. The nationwide standards allowed light trucks to improve their fuel economy more slowly than passenger cars. The physics was simple: heavier vehicles take more energy to move than lighter ones. Thus, increasing efficiency in trucks would be more expensive than for cars and was brought along more slowly. In 1978, the year the requirements kicked in, cars were required to get 18 mpg, followed by 22 mpg in 1981 and 26 mpg in 1984. By comparison, light trucks only had to get 20 mpg in 1984, the year the Dodge Caravan was introduced.

While this concession undercut the national effort in the 1970s to rapidly decrease dependence on foreign oil, its impact at the time was limited, because only 20 percent of the vehicles on the road fell into the light truck category. Additionally, these two separate catego-ries of vehicle were originally used for different purposes. Passenger cars were used for commuting, running errands, and driving kids to school. Light trucks, like pickups and vans, were primarily used on farms or by tradesmen. All this began to change with the Caravan.

Even though a minivan is essentially a box dropped on top of a car chassis, Chrysler managed to emphasize the utility-based "van" element and get the new vehicle classified as a light truck. As a result, a Caravan only had to get 20 mpg, while a station wagon released the same year had to achieve 26 mpg. The minivan, essentially a next-generation station wagon, extended the benefits of lower fuel efficiency to vehicles that were not used for commercial activities. By blurring the distinction between light trucks and cars, the Caravan solved their problem. In the next decade this distinction would be all but erased.

Chrysler executives also managed to effectively lobby the government to get its redesigned sport utility vehicle, the Jeep Cherokee, to be classified as a work truck for fuel efficiency purposes. The jeep had once been a very utilitarian vehicle; it was developed for use on World War II battlefields. But by the 1980s the vehicle was being increasingly marketed to suburban families. Regulators and auto executives alike were well aware of the yawning gap between the vehicles' official categorization and its real world uses. J. C. Collins, Ford's top marketer for SUVs and minivans said, "The only time those SUVs are going to be off-road is when they miss the driveway at 3:00 a.m."[2] But Collins's job was marketing, not fuel economy, and SUVs were a goldmine.

In 1990, six years after Dodge Caravan was introduced, Ford released its Explorer, a midsized sport utility vehicle. The Ford Explorer sold roughly 140,000 vehicles in its first year. In 2001, the truck's sales number more than doubled to 282,000, a success that signaled the explosion of the SUV market in the 1990s. Detroit rejoiced. Not only did each company have successful models—GM offered the Chevy Tahoe, Chrysler the Jeep Cherokee—but the profit margins on pickup trucks and SUVs were much higher than on cars. In 2000, Americans bought 445,000 Ford Explorers, a high-water mark for domestic auto manufacturing. The Big Three set consecutive annual sales records with 16.9 million vehicles sold in 1999 and 17.4 million in 2000.

This trend not only made a lot of money for the car companies; it put a lot more of gas-slurping trucks on the road. In the 1970s, when the distinction between trucks and cars was written into fuel economy laws, light trucks only made up about 20 percent of the national fleet. By 1984, about a quarter of vehicles were classified as light trucks. But over the 1990s, Americans developed a huge appetite for the space and sense of security that the larger, higher vehicles offered. By 2000, more than half of the vehicles on American roads were trucks—or at least they were considered light trucks by the government's CAFE standards.

The fleet mix had undergone a revolution, undermining the intent of the fuel efficiency rules. In 2004, light trucks were required to get only 20.7 miles per gallon. This was a full seven miles per gallon less than passenger cars. More shockingly, it was roughly equal to the fuel economy requirements that all cars already had to meet in 1980, over two decades earlier. The explosion of SUVs and minivans literally turned the clock backwards on fuel economy.

The new consumer appetite for bigger, less efficient vehicles had a similarly deleterious effect on the amount of carbon gases being emitted on American roads. Just as fuel efficiency is measured in miles per gallon, carbon dioxide emissions are measured in grams per mile. Less efficient vehicles burn more carbon dioxide-emitting gas to go the same distance than higher efficiency vehicles. As a result, when the mpg of cars dips, their carbon emissions shoot up. Over the first decade of the CAFE standards, carbon emissions dropped from an average of 590 grams per mile in 1977 to 405 grams per mile in 1987. But as more SUVs and minivans came onto the road, that number began a steady upward climb. By 2004, the new cars on America's roads were spewing an average of 461 grams per mile of heat-trapping carbon dioxide into the air, erasing years of progress on greenhouse gas emissions.[3]

But the SUV-driven records at end of the millennium turned out to be the top of the mountain. Sales began declining for the Explorer in 2001. The numbers for Ford Expedition and the Lincoln Navigator had also peaked. General Motors' leading models continued to top their sales numbers until early in the 2000s, when they began a steady retreat. This drop was partly due to fierce competition and the bursting of the Internet bubble that caused a recession early in the new decade. But it was also a sign of a larger shift in the marketplace. Call it the revenge of fuel economy.

★ ★ ★

The minivan and SUV had both stormed the market during an era of falling gas prices and relaxed CAFE standards. By 1984, when the Caravan was introduced, the oil scare of a decade earlier was a distant memory for most Americans. Without the consumer or the government to spur them upwards, American auto companies became more and more reliant on profits from their gas-guzzling trucks and SUVs over the course of the 1990s. But the cheap oil couldn't last forever. At the beginning of the new millennium, the price of gas began to climb back up, making those huge vehicles less appealing. In 2000, it would have cost around $40 to fill up the 28-gallon tank of Ford Expedition. By 2004 drivers were paying $75 for a fill up.

As gas prices went up, outside analysts and experts called for Detroit to invest in more fuel-efficient vehicles, but the Big Three refused to reverse course. They were addicted to the considerably larger profit margins from SUVs and pickups. In 2007, years after the SUV boom had slowed, trucks were still 70 percent of US sales for Chrysler and Ford, while they made up only 41 percent of Toyota's sales. Instead of reexamining their now less appealing fleet mix, the American companies directed blame for their losses and lost market share to management mistakes, overbuilt production capacity, growing pension and health care obligations, the supposedly unfair advantages of foreign competitors, and anything else they could think of.

For example, when Toyota surpassed GM as the world's leading automaker in the first quarter of 2007, GM blamed poor messaging about their increasingly out-of-step products. Chevrolet's longtime advertising agency was ordered to undercut the aura of Asian superiority in fuel economy. "You're not getting better quality [with Japanese cars]," said GM marketing chief Mark LaNeve, "you're not getting better performance, you're not getting better fuel economy—you're just paying a premium. We have got to get that communicated."[4] But the domestic automakers' decline continued.

In the spring of 2008, large SUV sales dropped 29 percent from the previous year. Meanwhile, sales of subcompacts shot up 33 percent, and the Toyota Prius hybrid grabbed a 23 percent increase. Midyear, Ford's F-Series, the three-decades-long sales leader in the United States, was being outsold by the Toyota's Camry and Corolla as well as the Honda Civic, all of them Japanese cars. Though the F-Series recovered its leadership position by the end of the year, its sales were down 25.4 percent from 2007.[5] But in the midst of their financial collapse, the problem facing Ford executives was bigger than their flagship model. The whole company was falling apart. Over the previous two years, Ford had lost over $30 billion. GM was in even worse shape, not having turned a profit since 2004. By the end of 2008, the giant had lost $74 billion over the previous four years.

Outside observers had been sounding alarms about this inevitable collapse for years. In 2006—when that twenty-eight-gallon tank cost $83 to fill—the *Christian Science Monitor* predicted that, "The marketplace will fix all this itself—eventually. Buyers now demand fuel efficiency, and the Big Three will shift gears, however painfully."[6]

That pain was now. Stockholder confidence in GM, Chrysler, and Ford plummeted, banks had no money to lend, and consumers were in no mood to buy new cars. GM predicted it would run out of money in the middle of 2009.

# 8

# GOLDEN STATE

Later, on February 2, after attending the Washington Auto Show, we have a meeting with Mary Nichols, the exuberant head of California's Air Resources Board. I love being in meetings with Mary because we can both confidently talk automobile technology in a room full of car guys. She has also done more to promote sustainable transportation than anyone else in the world.

Although she worked at the EPA for a number of years in the early 1990s—she was my boss and appointed me to my job at OTAQ—Mary would just as soon be in California. She has been an indefatigable champion of improving air quality during three terms as the chairman of California's Air Resource Board and later as the California secretary of the interior under Governor Schwarzenegger. *Time* magazine would name Mary as one of the world's one hundred most influential leaders in 2013. She is in town for the auto show.

We get the leader of her technical team, Tom Cackette, linked in via video from Sacramento. Tom plays the same essential role for Mary that Chet France does for me. Then I pass out two sheets to Mary and Lisa Heinzerling that read "EPA Staff Draft."

Back in California, Mary is overseeing implementation of the new greenhouse gas rules that President Obama asked us to reconsider last week. We want to brief her on our even bigger plans. A series of bullet points sketches out our strategy for taking California's aggressive greenhouse gas standards nationwide. This is not the first time the EPA and California have collaborated. Because of the state's unique history and role in cleaning up vehicles emissions, partnering with them on these rules is another essential tool.

★  ★  ★

Because it rose from an obscure port to an international city in the early decades of the twentieth century, Los Angeles was the first large urban area to be designed to the specifications of the automobile. While older cities like New York, Philadelphia, and Boston had relatively compact centers, Los Angeles was sprawling and laced with freeways almost from the beginning. Cruising on highways became such a common feature of life that a new term—surface roads—was coined to distinguish between highways and those smaller, local passageways that used to be simply be known as "roads." But as the city's love affair with cars continued unabashed during the twentieth century, an unwelcome side effect arose. In the 1930s and '40s a strange, yellowish brown haze began floating around Los Angeles and other urban areas in the state. The offensive fumes caused residents' eyes to become inflamed and teary, a burning sensation in the throat, and bouts of nausea and headaches. When the mysterious aerosol was thick, businesses would close up shop and children were kept indoors at school.

In extreme—but increasingly frequent—cases, the pollution nearly eclipsed the sun. Drivers had to flick on their headlights during the middle of the day, automobile accidents caused by pollution became common, and Los Angeles International Airport had to cancel flights. A new term, combining smoke and fog, was coined for this unexplained miasma, "smog." Though nobody could explain exactly where it originated, the people of Los Angeles knew they wanted it to go away.

In 1943, the *Los Angeles Times* ran a story describing public reaction to the phenomenon. "Everywhere the smog went that day, it left a group of irate citizens," reported the newspaper. "Public complaints reverberated in the press. . . . Elective officials were petitioned."[1] Los Angeles was hardly the only urban area in the country experiencing a smog problem. But the rapidly growing city was quicker to adopt aggressive measures to deal with the issue.

In October of 1943, a month after that article ran, Los Angeles County formed the Smoke and Fumes Commission, a group that rounded up the usual air pollution suspects while seeking the mysterious culprit. First they tried temporarily shutting down and limiting emissions from some factories, including shuttering a new rubber plant. Then they banned burning in open dumps between April 15 and November 15. Smog monitoring stations were set up around Los Angeles. In the early days of smog detection technology, these stations sometimes measured the fume's severity by monitoring the damage to the spinach and alfalfa plants kept on site. But none of these measures made a difference. In fact, the smog got worse—and it was intensifying in other urban areas.

In November of 1949, thousands of fans at a football game between UC-Berkley and Washington State experienced "intense eye irritation." Though much of the Bay Area experienced smog, these reports were very specifically focused on one geographic area around Berkley, and not a part of the region with a heavy industrial base, nearby garbage incinerators, or other known air polluters. In its report on the incident, the Committee on Air and Water Pollution focused on the one notable circumstance of the smog event: thousands of people packed in bumper-to-bumper traffic. Maybe, the commission concluded, the eye irritation was "in some way directly related to automobile exhaust." Could cars be the problem?

The idea that smog was somehow related to the process of burning petroleum was not completely novel. The Western Oil and Gas Association had financed a study on smog that pointed a finger at sulfur dioxide, a by-product of oil and gas combustion. Stringent anti-sulfur laws were immediately passed in Los Angeles County, which soon had the cleanest urban air—as measured by sulfur dioxide—in the country. But, despite the sulfur dioxide treatment, the smog was still getting worse. The state was making very real efforts to limit the smog clouding its cities, but it was clear a new approach was needed.

Then, one day in the mid 1940s, a Dutch researcher named Arie Haagen-Smit stepped outside his lab at the California Institute of Technology. Haagy, as he was known to his colleagues, was a chemist. He had a developed a technique that allowed him to isolate the chemical flavor components of natural foods like the pineapple. A muscular rower and avid outdoorsman, Haagen-Smit went on camping trips around the state to gather tree leaves to use in his experiments.

But the smell that greeted him on his break this day was the "stinking cloud that rolled across the landscape every afternoon."[2] Haagen-Smit felt like the air was attacking his lungs. He wanted to fight back. Since his biochemical technique involved skimming off and capturing the flavors and smells of foods before isolating them, Haagen-Smit figured he would do the same with the foul smell in the Pasadena air. Within a few years, he and a colleague had determined that smog was created by unsaturated hydrocarbons released from the gas tanks and fuel exhaust of cars.

Haagen-Smit's work formed the scientific basis for widespread acceptance that cars were a large part of the smog problem, but he didn't stop there. In 1950, Haagen-Smit turned from scientist to activist and publicly pressured for political solutions to the dire air pollution. Not surprisingly, his efforts were opposed and stonewalled by the car companies, who hoped to kill or delay any emissions requirements on their products.

In 1953, for example, the Automobile Manufacturers Association, the main industry trade group, formed a committee to investigate auto pollution. The work moved at a glacial pace. A year later, ten auto engineers flew from Detroit to Los Angeles to study the smog. After a week, the team reported that the automobile industry needed to do a "tremendous amount of work" before the pollution could be understood. The trade group promised that it would, "do whatever we possibly can to assist in the situation of automobile exhaust fumes' part in air pollution. We are dead serious." The automakers then spent years dragging their feet on any pollution mitigation measures.

Haagen-Smit was undeterred. In 1960 California established the nation's first motor vehicle emissions control board. In 1966, the state became the first in the nation to establish the "tailpipe" or exhaust motor vehicle emission standards.

In 1968, Haagen-Smit was nominated by then-governor Ronald Reagan to be the first head of the newly created California Air Resources Board or CARB. As director, Haagen-Smit was endowed with a power to regulate automobiles unparalleled in any other state. A few years after he had been confirmed as chairman, Volkswagen was late certifying its fleet's emission standards for that model year. Haagen-Smit immediately banned the sale of Volkswagens in the state. The company rushed its certification to him ten days later.

Even more important, the regulations that Haagen-Smit fought for went national. The Clean Air Act recognized California's nation-leading efforts to control its air pollution and gave the state the unique power to set its own automobile standards as long as they received approval from the EPA.

California's leadership was amplified by the 1977 amendments of the Clean Air Act. Not only could the state set stronger car standards than the national ones, but other states with significant air pollution problems could choose to institute the California requirements instead of the EPA rules. Allowing an agency to supersede federal authority in its own state was very rare. Giving it the power to set the agenda in other states was even less common. It was an authority that California has made good use of over the years.

The federal exception for car emission standards put California in a very powerful position. The state has both the largest population in the United States and a particularly strong attachment to its cars. California is far and away the single biggest car market in the country, and one of the largest in the world. Its sheer size, combined with the more stringent auto emissions goals it has pursued for the 27 million cars, trucks, and buses on its roads, often creates conflicting pressures for automakers.

When California passes tough new regulations, car companies can't simply refuse to comply and abandon the state's huge auto market. Their whole business model is predicated on producing the same style of car on a massive scale. Selling one model in California and another one in Texas is not as profitable. Automakers would much prefer selling the same model of car in California and the other 49 states. So, once California demands stricter standards, car companies have to weigh their options. One is to sue the state in hopes of killing or at least delaying implementation of the stricter rules. Another course of action is negotiating an agreement with California and the EPA. Since the late 1990s, automakers have offered to accept more stringent emissions standards in exchange for California's rules being harmonized with the rest of the country. If they can't avoid regulation altogether, the car companies would rather have the certainty of a unified, standardized national market.

All of this is on my mind during my meeting with Mary Nichols. Convincing her to bring in the Golden State as an EPA ally is essential to moving forward on our national plan. But not only was California's partnership an essential tool, our greenhouse gas plan had its roots in an extraordinary turn of events that played out a few years earlier in Sacramento, California's state capital.

★   ★   ★

In January of 2001 two California-based environmentalists, Tim Carmichael and Russell Long, walked into the office of a state representative named Fran Pavley. A freshman, Pavley had spent two decades as a public school teacher and a local politician in the small southern Californian town of Agoura Hills. She had long silver hair, a calm disposition and was just unpacking and beginning to find her way around the Assembly.[3]

Carmichael wore wire frame glasses and represented the Coalition for Clean Air, a California-based advocacy group that dated back to the early 1970s. Long had the relaxed smile of someone who has spent months at sea and was the founder of the Blue Water Network, a Bay Area environmental group. The men had an ambitious suggestion for Pavley. "What about a bill relating to tailpipe emissions to reduce greenhouse gas emissions?"

They had approached the assemblywoman after bouncing around ideas that would allow their state to do what the federal government had proved incapable of so far: combat greenhouse gas emissions. Since California had a leadership role and lots of latitude on auto emissions, why not treat greenhouse gases just like hydrocarbons, carbon monoxide, nitrogen oxide, and other auto exhaust pollutants? Regulating California's millions of cars for greenhouse gases would be a small but serious step toward addressing global warming. What's more, other states could opt into such a program.

Many more seasoned politicians wouldn't have touched such a bill. It had never been done before. It would be opposed by car manufacturers and auto dealerships. Even if the bill managed to become law, it would become the object of intense litigation. But Pavley, who had been carried into office in part by grassroots environmental support, was willing to introduce the bill to the new legislature.

Her proposition was immediately attacked by the Chamber of Commerce as a "job killer." But, perhaps because Pavley was such a low-profile member of the assembly and the bill was considered a long shot, the negative attention was relatively muted. This wouldn't last.

Over the rest of the year, Pavley's coalition slowly grew. The National Resources Defense Council and Sierra Club supported the bill, providing on-the-ground logistical support. Coincidentally, the Union of Concerned Scientists released a report detailing the impact of climate change on the Nevada snowpack, California's primary water source. These concerns prompted ski associations and water agencies to join Pavley. The report also helped Californians make the link between this global phenomenon and its local effects on their state: sea level rise, higher temperatures, and worsening air quality in the vast central valley. Soon non-environmental groups were also supporting the bill, including the politically powerful Latino caucus and a new group called Environmental Entrepreneurs, a collection of executives who put a business face on the effort.

A coalition of hundreds of Christian, Jewish, Muslim, and other religious leaders called Interfaith Power and Light believed that addressing climate change was a moral imperative. As the vote drew near, a priest visited a key legislator who was wavering on his vote. The bill passed with two Republican votes and moved over to the Senate.

Despite her growing ranks of supporters, Pavley's proposed legislation also benefited from her low-profile freshman assemblywoman status. She and her bill had been flying under the radar, remaining virtually nonexistent to many opponents. Once the bill made it to the Senate, this changed.

The car companies began spending millions of dollars to discredit Pavley's bill. Conservative talk radio began hammering the proposal for hours at a time. Soon listeners were angrily calling their state senators telling them to vote against the "anti-driving" measure. Auto dealerships took out full-page ads against Pavley's legislation. An astro-turfing campaign called "I want my SUV" warned that the bill would decide what kind of car Californians drove and force them into unsafe and more expensive smaller vehicles. The hosts of the *John and Ken* talk radio program organized a caravan of SUVs to drive the six hours from southern California to Sacramento, where they encircled the Assembly building honking their horns.

Inside, the corporate lobbyists and lawyers were also laying siege to the bill. During hearings, the chairwoman of the Senate committee reviewing the legislation became so sick of auto representatives threatening to sue that she finally yelled, "Send in your engineers, not your lawyers! We know you can do this!"

Pavley and other legislators also received death threats.[4] But she was a passionate environmentalist. As a teacher, she had seen the effects of air pollution in her students' many asthma attacks. Pavley didn't consider backing down. Fortunately, her bill was also adopted by John Burton, the tough-talking leader of the Senate. He bluntly told the Democratic caucus: "We're supporting this bill. You can remain neutral, but you cannot oppose it."

Having silenced dissent in his own party, Burton then dealt with the millions of dollars of advertising urging Californians to demand that their senators vote against "Assembly Bill 1058" with a legislative trick. Over a weekend, Burton took the entire contents of Pavley's bill and dumped them word-for-word into an unused bill numbered 1493. Overnight, "Assembly Bill 1058" disappeared. The flood of media and advertising encouraging people to demand their legislators reject a bill were attacking a nonexistent target. A few weeks later the bill passed the Senate. But it still faced its final legislative challenge back in the Assembly.

For Pavley, the days of mounting high-pressure politics were full of surprises. One day while sitting in her office in a building now constantly packed with lobbyists, Pavley looked up to see the new speaker of the Assembly, Herb Wesson, running into her office.

"Bill just called!"

"Bill who?" asked Pavley.

"President Clinton." The overlooked bill introduced by a freshman senator now had a national profile.

As the debate became more high-stakes and fierce, the car companies managed to scare off some of the assembly people who had originally supported the bill. Pavley needed three more votes. Then a political strategist from the Environmental Defense Fund named Jim Marston arrived. Marston was battled-tested from many years of fighting for clean air in the state of Texas. He had a distinctly thick Texas drawl and proudly wore his cowboy boots in California. Marston cloaked a cunning political sense in a disarming "down home" charm. His boss, EDF president Fred Krupp, send him to California with a simple directive: "You cannot loose." [5]

On the flight Marston studied intensive research on eight members of the assembly whose votes they might be able to get. By the time he arrived in Sacramento he had memorized their names, voting records, and districts as well as details like who their pastors were or the names of their best friend in college.

Soon after arriving at the capitol, Marston met John Burton. He introduced himself, explaining that he lived in Austin and worked out of the Environmental Defense Fund's Texas office. Burton quickly zeroed in on the fact that Marston knew Austin's congressman, Lloyd Doggett. Burton had been impressed by Doggett's strong opposition to a recent airline effort to "steal" money from the government following 9/11.

"Lloyd Doggett's got big balls," said Burton. "Have you got balls that big?"

"Well," said Marston. "I'd like to think they're big enough."

From then on Marston got strategy calls day and night from Burton. The Environmental Defense Fund was originally going to spend the hedge fund manager's donation on television ads to counter the enormous sums that car companies were pouring into their campaigns, but Marston didn't think that spending one million dollars on airtime was going to have much impact on the vote. The car companies were already spending a million dollars every week. Instead he focused intently on the eight assemblymen and women necessary to win.

Marston flew in Catholic priests to talk with assembly members about how important this measure was to the future of California. Actor Paul Newman called three female members. Two of them were over forty and thrilled to get a call from the movie star, but the younger woman didn't have any idea who he was. As the vote neared, Jim Marston counted four of the eight on their side—enough to get passage.

The morning of the vote, it was still unclear who had the edge. Speaker Wesson leaned heavily on an assembly member who had agreed to vote for the bill as a courtesy to a fellow legislator who couldn't be there. The absent assemblyman was home with his ill wife, but, under the immense pressure of the moment, his colleague was wavering on his promise. "If you don't vote for this," said Wesson, "we're going to tell him to drop everything, leave his sick wife at home, and fly him back up here."

In the middle of the proceedings, the Republicans called a party caucus and left the floor for three hours. Some Democrats opted to wait on the floor the whole time, because they were sick of being harassed by lobbyists in the building's packed hallways. During their meeting, two Republicans walked across the street to be guest hosts on the *John and Ken* conservative radio talk show. The show provoked another round of phone calls from listeners to legislators' offices. But when the voting finished, the first bill regulating greenhouse gas emissions had passed with a bare one-vote majority. The rookie schoolteacher, with help from environmental groups and powerful politicians, had shepherded through the first law limiting greenhouse gases in the country.

The law was ushered in with two signing ceremonies, one in Los Angeles and the other in San Francisco's Golden Gate Park. After that pageantry, the California Air Resources Board staff spent several years developing the state's regulations that were approved by the Board in September 2004. Finally, on December 21, 2005, CARB submitted to the EPA a comprehensive waiver request document for California to set its own greenhouse gas emission standards for light duty vehicles. Now, four years later, President Obama has ordered us to reconsider that same waiver, which was denied under President Bush.

With that order, my EPA team and I knew that the Pavley law, passed in 2002, was going to be the template under President Obama for the national program to regulate the greenhouse gases that had continued to build up unchecked during the rest of President Bush's presidency. The law had aggressive targets for greenhouse emissions reductions looking out into the future. Implicitly, it had the support of twelve other states—New York, New Jersey, Connecticut, Maine,

Maryland, Massachusetts, New Mexico, Oregon, Pennsylvania, Rhode Island, Vermont, and Washington—which had historically followed the standards set by California. Additionally, the governors of Arizona, Colorado, Florida, and Utah had said they would also follow California and the twelve states.[6] These seventeen states account for nearly half of all the vehicles sold in the United States. As Fran Pavley, who had traveled to some of the other states, said to me, "You could feel the momentum on the ground."

# 9

# TORTURED POLICY

On Thursday, February 5, 2009, three days after our meeting with California's Mary Nichols, China declares a state of emergency. The central government is alarmed by an extreme drought in eight northern and central wheat-growing regions. One of them, Henan, is larger than the state of Georgia and has gone 105 days without rain, the longest period in over half a century. Over the past decade, the drought frequency throughout a huge swath of northern China is the highest in nearly a hundred years. Increases in severe, extended drought worldwide are among the likely impacts of global warming.

On Friday, Sweden permits construction of the first nuclear reactor since a 1980 ban. Despite environmental concerns, the country sees it as a potential solution to the dangerous buildup of carbon dioxide emissions. Sweden isn't alone in judging the possibility of a local reactor disaster less of a threat than global warming worldwide. Finland is already constructing the first reactor built in Europe since the 1986 nuclear disaster in Chernobyl sent a radioactive cloud as far west as Ireland, eventually killed thousands from cancer, and dampened European enthusiasm for atomic power.

On Saturday, brushfires roll across the Australian state of Victoria. The conflagrations kill 173 people and burn power lines that link inland utility stations with the four million people in Melbourne.

Over the past several days, thousands were without power in the early evenings when temperatures reached a record 113 degrees and air conditioning usage peaked. The conditions are a reminder of the kinds of drought and heat waves climate scientists expect to become increasingly common in the coming years.[1]

That Monday morning, February 9th, I'm at my desk in Washington reading a message from Bill Charmley, one of the engineers at EPA's Ann Arbor lab. His email was sent at 11:38 a.m. and is so dry that it is hard to connect with the ongoing climate abnormalities worldwide. "Based on our recent discussions regarding the scope and time frame of a national GHG vehicles proposal, we have gone back and re-examined our approach and schedule. As you know, our technical efforts to date have been looking at time frames as far out as 2025. However, if we target our efforts on the development of a near-term (2012–2016) rule, I believe we could complete a proposal by the end of July or early August."

Bill's fairly wonky email on my standard-issue government black desktop PC is hardly a thrilling combination. But the technical details he is discussing are critical. Without the strongest science and policy behind the first federal rule to specifically combat global warming, we are likely to fail.

We've decided to pursue a rule that more closely follows the California legislation. Proposing a rule out to 2016 is a lot easier than creating rules that project further into the future, such as 2025. Based on current technology, we can reasonably assess what sort of greenhouse gas reductions automakers will be capable of in the 2016 timeframe. But that stretch of time still gives the automakers what they want: a predictable, unified national market that is guaranteed for years.

Normally, the rulemaking Bill is discussing would compel the EPA staff to do all sorts of intensive economic, policy, technical, and legal analyses. This process takes a minimum of two years, but it so happens that we already have a lot of solid science, cost analysis, and policy sitting on the shelf from the Bush years. That work came from the most optimistic era of the Bush administration. At the time, we still held out hope that our efforts might become the basis for action on greenhouse gases. Despite our enthusiasm, the thousands of hours of labor are still unused. We underestimated how far the Bush administration was willing to go to avoid regulating greenhouse gases. Our expertise today is a resource honed through years of frustration and pain.

★   ★   ★

By 2007, pressure to take some sort of action on climate change was building on the Bush White House from various sources. That year, a Supreme Court consisting of primarily Republican appointees issued its *Mass vs. EPA* decision and ordered the EPA to determine whether or not greenhouse gases from motor vehicles constituted a threat to the public. That mandate came from the highest court in the land; there was nowhere to appeal it. Unless the Clean Air Act was rewritten, the president had no other options.

On the other side of the country, California Republican governor Arnold Schwarzenegger had been waiting two years for a response on his state's request for a waiver to institute the Pavley greenhouse gas program. All the state needed was the EPA to review and grant their request. The agency had approved every single one of the state's forty previous requests.

So, on May 14, 2007, President Bush did respond to the Supreme Court's order by signing Executive Order 13432. As its numerical name suggests, this was only one of dozens of such documents that presidents issue every year. The previous Friday, Executive Order 13431 had established an Iraq Transition Assistance Office. On Friday the 18th, Executive Order 13433 prevented federal agencies from hiring lawyers or expert witnesses on a contingency basis. But Executive Order 13432 had a special significance to us because it required the EPA to determine whether or not greenhouse gases were a public danger. It was an opportunity many of us had been looking forward to for years, but it wasn't as straightforward as it could have been. By allowing EPA to prepare a report on greenhouse gases, the president had complied with the letter of the law. But he wasn't required to make it easy. In fact, the administration made sure that the bureaucratic maze in which we worked was by turns convoluted, hostile, and obstructing.

For starters, the Supreme Court's order was directed at the EPA. But the executive order required the agency to work in tandem with the Department of Energy and NHTSA. Neither was likely to be aggressive in regulating greenhouse gases. The Department of Energy was led by Samuel Bodman, who was on record as questioning the link between greenhouse gases and global warming. But we'd be working much more closely with NHTSA, the gatekeeper of any increases in fuel economy.

Because of decades of inaction on fuel economy and budget cuts, NHTSA was understaffed and relied heavily on the car companies for the technical information on which they based fuel efficiency standards. But the potentially biased numbers were not the biggest threat to effective standards. It also turned out that John Graham, the head of the Office of Management and Budget, was quietly trying to dictate to NHTSA staff the type of CAFE standards the White House wanted to see.[2]

As a result, NHTSA had a difficult time developing strong fuel economy standards. In fact, six months after President Bush issued his executive order to examine the public dangers of greenhouse gas, a number of environmental groups and other advocates for public health successfully sued NHTSA for its weak regulation. The plaintiffs claimed that NHTSA's new fuel economy standards were too low. The court agreed and ordered the agency to revise their numbers.

So, while President Bush's May 2007 executive order did compel the EPA to fulfill the Supreme Court's demand about greenhouse gases, it also forced us to work with two historically less aggressive partners. Worse, the order set the stage for a possible turf fight. NHTSA's territory was fuel economy; the last thing they wanted was another agency moving in with a new criterion—greenhouse gases—on emissions. From the EPA perspective, NHTSA had none of our in-house expertise dealing with the science and health issues of greenhouse gases from auto emissions. This was strictly EPA territory.

The potential for a rocky working relationship certainly wouldn't bother anyone at the White House. Any tension would likely make sure EPA didn't move fast or get far in responding to the Supreme Court's order. Unfortunately, our difficulties with other agencies were not nearly as big an impediment to effective work as our own boss, EPA administrator Steve Johnson.

★   ★   ★

Steve had been one of us. After joining the agency in 1979—the year before I did—he had spent two decades as a career EPA scientist. Then, after quietly rising through the ranks of civil servants, he took a political position as the lead toxics and pesticides official in 2001. Soon Steve was appointed to deputy administrator and then on to the top spot as EPA administrator in early 2005. As with almost all Bush appointees, he displayed a willingness to walk lockstep with the administration.

Steve was a tall, affable man with silver hair that sat parted and flat across his head. The senior staff he invited to his large, high-ceilinged office found him smart and engaged. He held a degree in biology, asked good questions, and clearly understood the details of his agency's work. He invited people to speak their minds. But all his openness ultimately amounted to nothing. Despite having a grasp on the facts his staff brought him, Steve's final word on important EPA policy was, I felt, more often than not influenced by people working down the street at the White House.

During my first meeting with him, an annual strategy session for all the agency's senior executives, Steve kept talking about his recent visit to Camp David. "'Can you believe it?' I said to my wife. 'Here we are with the president!'"

Unfortunately, it was also one of the few times he would have that honor. Throughout his tenure, Steve had very little personal access to

President Bush, or the White House generally. The EPA administrator is generally considered a cabinet-level position, but Steve was working back channels to get the president's attention. Strangely, one of his closest allies in this effort was Hank Paulson, the secretary of treasury. Although he had publicly supported the president's positions, Paulson had become concerned about the potential economic impacts of climate change. Steve relied on him to bring his messages to President Bush's attention during their weekly meetings. Paulson was sincere, but, by late 2007, it's hard to imagine that these discussions didn't almost exclusively focus on the collapsing economy.

★   ★   ★

Despite Steve's often politically driven leadership, we produced a scientifically sound response to the Supreme Court's order before the end of 2007. Under the leadership of Dina Kruger, director of EPA's Climate Change Division, the staff determined, and Steve concurred, that greenhouse gases were indeed a danger to the public welfare. In other words, carbon emissions from cars, trucks, planes, and utilities were a threat to Americans.  On the morning of December 5, 2007, we submitted our exhaustive research and what was known as a "positive endangerment finding" to the White House's Office of Management and Budget (OMB). These types of regulatory reviews have been in place since the early eighties when President Reagan first ordered them. But even this procedural step wasn't made easy. In fact, it was impossible.

At about ten that morning, Jason Burnett, an EPA senior advisor to the administrator, called Susan Dudley. Susan's position as administrator of the Office of Information and Regulatory Affairs with the Office of Management and Budget made her the highest-ranking regulatory official in the executive branch. She was the gatekeeper to the White House. She was also married to Brian Mannix, the EPA's associate administrator for the Office of Policy, Economics and Innovation. They were both formerly of the conservative Mercatus Center at George Washington University. Neither was a fan of regulation.

"We're getting ready to send over the positive endangerment find-ing," said Jason.[3]

"Ready to go," said Susan.

Jason emailed her the large file containing over 4,000 pages of our scientific analysis. Five minutes later, EPA administrator Steve John-son got a call from the president's deputy chief of staff, Joel Kaplan. "Don't send the endangerment finding," Joel said. "There is discussion over at the White House about changing course over the pending legislation."

The "pending legislation" was the Energy Independence and Security Act, a Democratic-sponsored bill that focused on automo-bile fuel economy, the development of biofuels, and energy efficiency in public buildings.

Steve walked down the hall to Jason's office.

"Did you send the endangerment finding?"

"Yeah. Five minutes ago," replied Jason.

Steve walked back to his office and called Joel Kaplan. "Sorry, we already sent it."

"Can you say that you sent it in error?" asked Joel.

Steve passed on the message to Jason.

"Well, I didn't send it in error," said Jason. "So I'm not going to say that I did."

Steve was generally happy—even honored—to follow the presi-dent's orders, even when they were obviously driven by policy and not science. But now he was being asked to admit to an error he hadn't committed. Steve called back over to the White House. "Joel, we can't do that."

"OK," said Joel. "Can you retract it? Say you had second thoughts?"

"Steve, do you have second thoughts?" asked Jason.

"No."

"Neither do I."

Joel responded that the White House didn't plan to review the email.

"Well, I've already sent it," said Steve.

"As a practical matter," said Joel, "we believe that if we don't open the email, we don't believe we are in receipt of it."

And that was it. Thousands of hours of work to produce a well-documented and scientific response to a Supreme Court ruling and an executive order sat unopened in a White House email account, probably for the rest of the Bush administration. One of the top officials at the White House had essentially jammed his fingers in his ears and pretended not to hear our warnings on greenhouse gases. It was an extraordinary political effort to suppress science that the administration didn't like, but it was not the last insult to EPA staff.

★   ★   ★

Though the White House made our finding on greenhouse gas disappear, we had still moved forward with Governor Schwarzenegger's waiver. On June 7, 2007, I brought a handful of my staff to meet with Steve in his office. It was a short meeting. Karl Simon, one of directors responsible for the California waiver review, presented eighteen PowerPoint slides summarizing thousands of pages the staff's research on California's request to initiate their regulation of greenhouse gases in auto emissions.

Karl explained that under the law, California had to meet three tests to qualify to write its own laws. It had to show that its rules were at least as stringent at the federal regulations, that they were not arbitrary, and that the state had a compelling air pollution concern. As it had in every previous waiver request, California met all three tests. Then he told Steve and his staff that our analysis showed that the agency was required by law to grant California its waiver—and that we had never turned down any of their previous forty or so waivers. John Hannon and his legal colleague, Michael Harwich, indicated that if we denied the waiver and California sued, the EPA was "likely to lose," but if the EPA allowed California to proceed and automakers sued, that EPA "is almost certain to win."

Steve and the acting assistant administrator for the Office of Air and Radiation, Bob Meyers, listened and asked a few questions. Bob was bearded and usually wore a serious expression. We knew that he had been to the White House to talk about California earlier in the week. We didn't know what he had been told, but it was unlikely to be good news for my staff and me.

The following day, I got an email from Bob's chief of staff about a slide from our presentation showing that the EPA staff's initial assessment was that the waiver needed to be granted.

"Bob was not happy when he read that page during the briefing. He wanted to let someone at OTAQ know about this so we can permanently delete the offending language and not let it arise again." It wasn't much feedback. But it was really all we needed to know. No matter how rigorous our work, politics would have the final say.

On October 30, 2007, we had a final briefing with Steve in his narrow bullet-shaped conference room. Even before we walked in the room, the political press had started. The day before the briefing, Bob rewrote our legal assessment that waiver approval might mean a likely unsuccessful lawsuit by the car companies, while denial would almost certainly result in a successful suit by California. Bob told the staff to soften this legal warning to read "litigation risks significantly higher than if a waiver is granted." As I walked into the room, I saw that someone had set up a big screen at the back playing muted scenes from the campy movie *Batman & Robin* in which the current California governor, Arnold Schwarzenegger, played the villain, Mr. Freeze. In the background, the Eagles song "Hotel California" was playing. Steve wasn't amused.

Karl Simon outlined the staff's research, finding that, as it had in every previous waiver request, California met all three tests. The only real legal option was to let the state move ahead with its rule. Steve went around the room and polled everyone. All but Bob Meyers, who didn't offer an opinion, said we should grant California's waiver. Then we filed out and went on to our other work while Steve, and whoever else, made his decision. Meanwhile, something unusual had been going on in Congress.

★   ★   ★

Supreme Court orders and requests from California weren't the only factors that made action on climate change seem politically feasible for the first time in decades. Over the course of 2007, average gas prices in the United States had continued their upward climb, spiking at above $3.50 a gallon and causing customers to ditch their gas-guzzling SUVs. Toyota surpassed GM as the world's largest automaker. Years of expensive and bloody wars in oil-rich countries were wearing on the public. And the 2000s were on their way to becoming the hottest decade in United States history, surpassing the previous hottest decade, the 1990s. This confluence of events brought together an unlikely coalition of senators and car companies to support what became known as the 2007 Energy Independence and Security Act, or EISA. Originally introduced by Democrats, the bill proposed a range of initiatives including increasing the volume for renewable fuels and raising fuel economy for cars and light trucks.

One of the most surprising supporters of the fight against climate change was Alaska senator Ted Stevens. A longtime critic of fuel standards, Stevens had often railed against so-called environmental extremists on the Senate floor. But a year earlier Stevens had visited the native Yup'ik villages in a remote western part of his state, a trip that brought him to the frontlines of global warming. He could see villages melting, homes and a way of life literally disappearing in front of him. The gruff seventy-three-year-old Republican had a conversion experience. He returned to his post on the Senate Appropriations Committee dedicated to doing something about climate change.[4]

From the other side of aisle came Byron Dorgan, a Democrat from North Dakota. Dorgan had been a longtime opponent of the standards as well, but had become concerned about the effect of America's dependency on foreign oil. He had begun to look at reducing gasoline usage as a national security issue and brought along Larry Craig, the Republican from Idaho, to support the 2007 EISA bill.

With rising gas prices, fuel efficiency was more important for American drivers than it had been in three decades, but the sudden prominence of that issue wasn't the only reason senators began embracing the bill. Among the new supporters was Trent Lott, the conservative Republican from Mississippi. Lott had once made an anti–fuel economy speech standing next to a purple subcompact. He theatrically claimed that federal regulations would force Americans into tiny, dangerous cars like the "purple people eater" next to him. But now Lott was also coming to the fuel efficiency table. Among other things, he had to consider the support of Nissan, whose assembly plant made them the second-biggest private employer in Mississippi. As it became more likely that the 2007 EISA would be passed, creating new CAFE fuel economy standards, Nissan didn't want to be opposing the law but wanted instead to give input into its final shape. Lott became another previously unimaginable supporter of the bill.

Even more surprisingly, Lott offered to bring more Republican votes with him. Senate watchers knew that Lott couldn't just pull in these votes from other states to support his state's Nissan manufacturing plant without there being some other motivation. These senators, and pretty much the entire Republican Party, had spent years opposing increases in CAFE fuel economy standards. For these senators to come around, the Bush administration must have been reevaluating its policies.

The tide in the White House was also changing. Back in 2006, President George W. Bush's press secretary, Ari Fleischer, answered a press query about whether President Bush believed in fuel efficiency standards for automobiles, saying, "That's a big No. The President believes that it's an American way of life, and that it should be the goal of policy makers to protect the American way of life. The American way of life is a blessed one. And we have a bounty of resources in this country . . . Conservation alone is not the answer."[5] More than likely, the White House had taken in the changing landscape— the Supreme Court order on greenhouse gases, the California waiver request, and the support for raising fuel efficiency for national security reasons—and decided that the legislation in Congress was the lesser of evils. In another time and place, the administration could have simply stalled and killed a CAFE increase. But an increase in fuel economy was much better than the possibility that the EPA and California might be allowed to regulate greenhouse gases. Given the options, the Bush administration would settle for a bill that increased fuel economy standards.

On December 18, 2007, the Senate passed the Energy Independence and Security Act, otherwise known as EISA. It was the first law to raise CAFE standards in over two decades, but I didn't spend too much time celebrating. Two weeks earlier, the White House's Office of Management and Budget had effectively killed our finding that greenhouse gases posed a danger to the public by refusing to open an email. I suspected that the new law was a tool the administration hoped they could use to kill the California waiver request. What was incredible was how immediate and transparent their effort was.

★ ★ ★

The next day, Steve called me into his office. He had just returned from China and seemed refreshed and excited from his time away. He also told me he had a letter to send to Governor Schwarzenegger, the long-awaited response to California's waiver request.

About an hour later, Steve, Bob Meyers, and I took the elevator downstairs and walked over to a Chevy Suburban. We were headed to the Department of Energy, where President Bush was announcing the passage of the Energy Independence and Security Act. Steve opened door for me. As I ducked into the SUV's huge interior, he said, "Now that we have the EISA law, we don't need the California waiver."

"Steve," I said, "the two are not related." The Clean Air Act requires that EPA evaluate the California waiver based on specific criteria. EPA's responsibility under Clean Air Act was not simply negated if another agency had a strong fuel economy rule.

"We'll talk about it when we get back to the office," responded Steve.

Two hours after returning from the announcement ceremony, I got a call. "Steve needs to see you. Come downstairs right away."

I went down to the bullet room and found Steve, Brian Mannix, Jason Burnett, and Bob Meyers along with Steve's spokesperson. Steve was stone-faced.

"Our press people got a call from the *Washington Post* that the information from the California waiver had been leaked," he said. "Some of the briefings have been released, including the one laying out the chances of EPA losing a lawsuit if we don't grant California its waiver."

Steve was not talking about the version with softened language, but the unredacted one that EPA lawyers John Hannon and Mike Horowitz had put together. They had put the likelihood of EPA losing a lawsuit with California at about 90 percent. Bob Meyers had tried to eliminate these unfriendly numbers from the official decision making process. Now the *Washington Post* was going to run a story featuring this analysis the following day. The media exposure would be taken as more evidence of the politicization of climate science in the administration. Steve was very nervous.

Steve and his team decided to try and head off any potential damage from the leak by putting together a letter laying out why they had rejected the EPA waiver and sending it to Governor Schwarzenegger immediately. He pushed it over his desk toward me.

"Steve, this is almost exactly the same thing you said on the way to the president's announcement. 'We denied the California waiver because of EISA.' But these two are unrelated."

I sat back. "We need to get John Hannon downstairs."

By now John was used to being put in sticky political situations, like when he had been assigned to defend EPA's position on the *Mass vs. EPA* Supreme Court case. He came down quickly and revised and softened the letter before it went off to California. But while the language was smoother, the facts could not have been clearer: the White

House wanted to use the first increase in fuel economy in two decades to kill action dedicated to fighting climate change.

★   ★   ★

At the beginning of 2008, morale at the EPA was very low. In the past month we had seen our work on the dangers of greenhouse gas disappear digitally. That 4,000-page endangerment finding was still sitting in political limbo as an unopened email. As a result, the government had still not responded to the Supreme Court's order. Weeks later, the administrator had overruled every single one of his senior career staff and, for the first time in almost forty years, denied California's waiver request, unlawfully. Meanwhile, our greenhouse gas regulation for cars in response to President's Executive Order was not moving forward. In the face of intense pressure for some sort of action on climate change, the Bush White House had managed to keep a tight lid on our efforts. Whatever it took, the administration was determined to tough it out for one last year.

Surprisingly, it was Steve Johnson who initiated the EPA's next climate change effort. On January 31, 2008, Steve called up Deputy Chief of Staff Joel Kaplan and asked him to pass along a plan for the EPA to explore greenhouse gas regulation to the president.[6]

"No," said Joel. "Don't send the letter."

"Joel, the president appointed me to be head of EPA. I need to tell the president what I believe."

"Don't send it to us."

"Well, it's coming your way."

This time, to avoid any digital tricks, Steve had the letter hand-delivered. Joel received it against his wishes.

Steve quietly wanted to do something in response to the Supreme Court mandate, so he sent a letter to the President Bush arguing that climate change is real and that the EPA has to release its endangerment finding and regulate cars for greenhouse gas emissions. Unfortunately, he had no personal access to Bush. The letter was accompanied by handwritten note for Joel to pass to the president. It is unclear if Bush ever saw the letter.

The letter became public when Congressman Henry A. Waxman released it. Many in Congress had become concerned about the recent developments, and California Congressman Waxman was one. He was not only the head of the powerful House Oversight and Government Reform Committee but also one of the most influential lawmakers, who for four decades led the legislative efforts to cut air pollution and reduce smoking. I had worked with the congressman and his staff since our 1993 second-hand tobacco report and later when I became the director of OTAQ.

In December, after Administrator Steve Johnson's denial of the waiver, the congressman investigated the decision to stop California's efforts to reduce greenhouse gas emissions from automobiles. On May 19, 2008, after months of investigation and a review of 27,000 pages of documents, Congressman Henry Waxman's House Committee on Oversight and Government Reform concluded that the White House improperly intervened in EPA's decision to deny the waiver.

Steve later initiated a report on how EPA might regulate greenhouse gases from all sources, from cars and trucks to power plants and agriculture. We hoped this effort—called an Advance Notice of Public Rulemaking (ANPRM)—would finally get our findings into the public's hands and salvage the work we had done since 2007. But, once again, we quickly encountered political manipulation. Among the sections stripped from the ANPRM report was a major portion from our proposal for new car rules and the conclusive evidence that elevated levels of greenhouse gases may reasonably be anticipated to endanger public welfare.

★ ★ ★

On July 9, 2008, a little over half a year since Steve had denied the California waiver and the White House had made our endangerment finding disappear, Brian Mannix strode into my office.

"I'm going to set up a meeting with Steve Johnson and the entire staff," the political enforcer Mannix said. "Will you arrange for the staff to give him a standing ovation when he arrives?"

I stared at him for a moment. He maintained a straight face. I stayed quiet.

"Steve's been advocating hard for you with the White House," Mannix said. "He needs the boost. The staff really should appreciate him more."

Because Mannix was married to Susan Dudley, the White House gatekeeper who had "lost" our endangerment finding and stymied everything else the EPA staff was working on, his request felt a little like blackmail.

"Steve is fighting hard for our greenhouse gas rule. He doesn't get the respect he deserves from the EPA career people." I still couldn't think of anything worth saying.

"Margo, you're respected by most of the staff at EPA. They'll listen to you. Can you arrange it?"

"Brian, I'm flattered that you think I have such power, but I can't deliver this," I told him. "The staff is discouraged and disappointed. Steve has betrayed us more than once. He has no respect for the tireless work we do, for the product we put out. He has no respect for the science and the law." I paused. "When we produce a strong greenhouse gas rule, Steve can back it with everything he has. Then he might earn a standing ovation."

Despite all of our setbacks, the Advance Notice of Public Rule Making on greenhouse gas regulation was one thread of hope. Steve still had a chance to gain some of our respect. But the day after Brian Mannix made his request my worst fears were realized.

Once it is finished, every ANPRM proposal goes through an interagency review. The other departments in the executive branch share their ideas and opinions. The Energy Department might have specifics concerns about the impact on utilities or the Department of Transportation might share specialized knowledge on highway congestion. Like a peer-review in academic circles, the interagency review will, in theory, create a stronger rule with broader appeal. After the process, EPA could decide which comments were useful and which ideas needed to be incorporated before releasing their final proposal.

The ANPR was formally published on July 11, 2008. However, instead of absorbing the mostly critical comments from other agencies, Steve decided to preface EPA's nearly six hundred pages of work by publishing the other agencies' criticism for the public to see. It was like putting bad reviews on the front cover of a book. Our own administrator had undermined our work before anyone even got to read it. The decision was described by the *Wall Street Journal* as an "extraordinary move." To us, it was another slap in the face.

The criticisms prefacing our document all began pleasantly enough. In her letter, the Office of Management and Budget's Susan Dudley called climate change possibly the "most significant environmental policy decision" of the day. In their combined letter, the Secretaries of Agriculture, Commerce, Energy, and Transportation stated: "Climate change is a significant issue." Then the letters switched to language like "complexity" that threw water on climate change action by claiming there was too much that was unknown. Not surprisingly, these responses were neatly in tune with decades of oil industry rhetoric: climate change was a "significant" problem but now wasn't the time to take action on it. Instead, greenhouse gases should accumulate at their present rate until some future "workable and meaningful" strategy came along. It was a flashback to fourteen years earlier when the tobacco industry launched an effort to discredit and distort our report on the dangers of second-hand smoke.

Unfortunately, this wasn't the worst of it. In the administrator's preface, Steve had not even paid lip service to the problem of climate change. Instead, he did everything he could to separate himself from his staff's work. The Clean Air Act, he claimed, was "an outdated law originally enacted to control regional pollutants that cause direct health effects, is ill-suited for the task of regulating global greenhouse gases." Any solutions based on the Clean Air Act would "inevitably result in a very complicated, time-consuming and, likely, convoluted set of regulations."

Our work was dead on arrival. It was an unbelievable ending to our years of fruitless labor. The *Washington Post* called his release of our proposal for a greenhouse gas rule "tortured policy."

But, now that Obama is in office, we have a chance to prove Steve wrong. We can finally use the reams of work ignored and suppressed by the Bush-era EPA as another tool to advance our goal of setting greenhouse gas limits. Between the Clean Air Act, the expertise of our lab, California's program to reduce greenhouse gas emissions from cars, the *Mass v. EPA* decision, and our work from the last administration we have a lot of formidable tools. We also have in our favor the executive orders signed by the president to address greenhouse gas emissions from cars by reevaluating California's waiver request, support of our new EPA administrator, the weakened bargaining position of the now crippled car companies, and increased consumer interest in higher fuel economy after a decade of rising gas prices. But before we can start negotiating with the automakers, we need to win over NHTSA, the Office of Management and Budget, the Council of Economic Advisors to the President, and the many agencies that review EPA's regulations.

## 10

# THE MOST POPULAR PEOPLE
# IN THE BUILDING

It's 6:12 p.m. on February 12, 2009, just over two weeks after President Obama has signed the executive orders. I am sitting at my desk, using two fingers to hunt and peck responses to a pile of email. A message from Lisa Heinzerling is headed: "You guys are the most popular people in the building."

She has forwarded a bunch of questions from Jody Freeman, a Harvard lawyer working out of climate czar Carol Browner's office at the White House. Jody is trying to get some of the basics about the potential for combining both NHTSA's CAFE fuel economy rule and our proposed greenhouse gas standards.

"I probably don't even know half the questions I have—lots to do with modeling and analysis relevant to setting mobile source standards under Clean Air Act assuming endangerment finding; how they compare (or don't) or convert (or don't) to fuel economy standards, and where points of disagreement/tension are on these issues. Trying to make sure I understand all the car stuff and know what everyone thinks from their perspective."

Jody's signature at the bottom of the email identifies her as the counselor for energy and climate change, which makes her my office's

primary link to the White House. Though the president's two executive orders allowed us to consider greenhouse gas regulations and
higher fuel economy on vehicles, the Obama administration hasn't
embraced any one specific set of rules. Our job is to get back to them
with a regulatory package that looks promising. Until then, we'll
have to contend with people inside the White House who may not
be thrilled about the prospect of placing new environmental rules on
the virtually bankrupt auto industry. We need their support before
we get into serious negotiations with the car companies. The backing of the new administration will strengthen our bargaining position considerably. Jody could be a very important advocate for us. I
write back to Lisa that it would be important to get to know Jody
and meet with her in person instead of trying to explain a bunch of
technical details via email.

Jody strides into my office the next morning. She is tall, polished,
unreadable, and originally from Canada. It turns out that she and Lisa
already knew each other from the university law professor circuit.
Jody is herself an accomplished environmental lawyer. She authored
an amicus brief for the *Mass vs. EPA* Supreme Court case and the
founding director of the Harvard Law School Environmental Law
and Policy Program. But though she was well versed in the legal
aspects of environmentalism, Jody was now immersing herself in the
details of vehicular engineering and fuel efficiency. She also had to
dig into the specific legal possibilities and contradictions between the
three different laws critical to our rule: California's state law to reduce
greenhouse gas emissions from new cars, which still requires a waiver
signaling EPA's approval; the Clean Air Act, which defines the regulatory parameters of federal emissions standards; and the EISA law,
which forms the basis for the federal CAFE standards. Finally, there
was the thorny question of how much any regulations would cost.
Lisa Heinzerling; my counsel, John Hannon; my senior policy advisor,
Maureen; and I sit at the conference table in my office. Chet, members
of his Ann Arbor team, and my deputy, Chris Grundler, are assembled
on the screen.

Jody gives a tight smile, sits down, and launches into a nonstop series of questions. She keeps rephrasing potential scenarios about technical and legal issues on fuel economy.

I respond to her questions about California's greenhouse gas regulations, and then say, "Jody, we have a plan that I have already shared with Lisa Jackson and Mary Nichols. It will allow for a national program to regulate both greenhouse gas and fuel economy through the next decade. We have already addressed virtually all the technical questions you are asking."

Then I ask Chet to go over our national program concept that regulates both fuel economy and greenhouse gases. Jody gives me a puzzled look. We spend the next hour going over her technical and legal questions. I am unclear if Jody is convinced that we have a viable path to a national car program, but she has stopped asking questions and is listening intently. Finally, I suggest that she talk to her boss at the White House, Clinton's EPA administrator, Carol Browner. Carol knows my office's work well from her days as President Clinton's EPA administrator and has already been briefed on the plan by Lisa Jackson and Mary Nichols.

After the meeting, Maureen suggests we invite Jody out to Ann Arbor to show off our facilities and experts. The lab is our trump card. Jody agrees to visit on February 26.

★   ★   ★

Two days after the meeting with Jody, an article in the *Wall Street Journal*[1] quotes Carol Browner saying that EPA will soon release a ruling that carbon dioxide represents a public danger. Finally, our long awaited response to the Supreme Court order that the Bush administration had sat on for the past two years is back in play. The article then goes on to quote the reaction of business groups that forecast the negative impacts on the economy of carbon dioxide regulation: William Kovacs, a vice president of the US Chamber of Commerce, warns that, "Once carbon dioxide is regulated, they can no longer contain the Clean Air Act . . . and it would completely shut the country down." Other business interests express fear that the ruling could "prevent power plants to operate and drastically increase the costs of power generation." Car makers piled on, expressing concern not only about the costs of meeting the tough new standards but also having to make cars that would have to meet two different mandates, a federal standard and a California standard.

Ever since the catalytic converter, we've been faced with the same kneejerk reactions. At the faintest scent of additional regulation, groups like the Auto Alliance immediately predict an industrial Armageddon.

★   ★   ★

On February 26, 2009, I fly Delta out to Detroit with Jody and her assistant, Jake, Lisa Heinzerling, and Transportation Secretary Ray LaHood's chief of staff, Joan Deboer. While we are in the air, General Motors announces that its cash reserves were down to $14 billion by the end of 2008. The company lost $30.9 billion in 2008 alone.[2] GM CEO Rick Wagoner meets with President Obama's Task Force on the Automobile Industry, and the company says that it cannot survive much longer without additional government loans.

Chet and some of his staff meet us at the lab's office. Tom Cackette, the deputy director of the California Air Resources Board, has also flown out to Ann Arbor for the visit. He doesn't say much, but his presence points to the strong relationship between EPA and California. About twenty of us pile into the lab's largest conference room.

One of our senior engineers, Jeff Alson, who has worked on climate and fuel economy issues for many years, steps to the front of the conference table and explains in scientific detail how a third of greenhouse gases released around the world come from transportation. Scientists, Jeff says, estimate that we need to reduce greenhouse gas emissions by 50 percent in 2030 and 80 percent in 2050 from 2005 levels to avoid increasingly high temperatures, volatile weather, and eventual catastrophe. At this lab we can accurately evaluate the technologies and the costs of setting greenhouse gas limits for cars, trucks, and other vehicles as well as the impact on climate change.

Jody spends the entirety of Jeff's presentation on her Blackberry. Chet keeps glancing at me, but there's nothing to do except move forward with our briefings. The presentation goes on for two and half hours. Jody eventually puts down her phone. I don't want to simply inform Jody on the questions at hand. I want to overwhelm her with our supremacy on the details of technology, the associated cost, and the résumés of our accomplished staff. I need to make clear the thousands of hours of work behind our proposed greenhouse gas plan and how we developed our technical capacities over the past decades of regulation.

Jody has a limitless appetite for information, but after an hour and a half I begin to wonder if the demonstration is a bit numbing on her. She is attentive but noncommittal. Is our show having its intended effect?

Afterwards, it's time for fun in the lab. We get Jody to climb into a car parked inside a testing cell on the huge dynamometer. Multiple hoses and wires attached to all sorts of different sensors and control systems snake out of the car. Though it's officially for emissions testing, driving it is a bit like playing a video game. Jody has to follow a route on a screen, first driving slow, then fast, accelerating up hills, and in stop and start traffic, all while not driving off the road. Engineers in an adjacent testing cell separated by glass peer at their computers as the values come in.

There are about twenty cars being tested in different cells on these huge treadmills. They run from a blue Ford Focus to a red Lamborghini. In other cells, engines from boats, bulldozers, and weed whackers are tested for pollutants. Engineer Joe McDonald demonstrates how EPA engineers retrofitted a Ford Expedition SUV with advanced catalysts to have 99 percent cleaner emissions five years before the regulation's deadline. Byron Bunker, another engineer, shows how our team developed technologies to reduce emissions by 95 percent and erase the black smoke coming out of a huge Cummins diesel engine. Charles Grey, who has been with the Ann Arbor testing lab since its inception in 1971, displays his newest invention, a hydraulic hybrid truck that could reduce commercial vehicle emissions by over 50 percent and double fuel economy. As we are winding up our tour of the lab, Jody's face relaxes a bit. Is she is making a connection between our technical ability and policy proposals?

# II

# VOLPE TO OMEGA

That Saturday, February 28, 2009, I receive an early morning message from Ron Medford, the head of the National Highway Transportation and Safety Administration. "I know everyone is very busy, but even if we are able to have our staff talk (by video conferencing) I think it will be helpful for us. Have a nice weekend! Ron." Ron always wears suits and is very cordial, but the relationship between our teams is strained.

Chet and his team have developed our own model to predict the cost of reducing greenhouse gas emissions and increased vehicle efficiency. Our model is called OMEGA. Not surprisingly, it produces very different results from NHTSA's VOLPE model. This is not helpful for responding to the questions Jody and other White House economists and lawyers are starting to ask us about the program.

Our model, OMEGA, allows us to estimate fuel economy improvements by considering how multiple upgrades might work in conjunction with each other. Making a vehicle lighter, turbocharging it, taking it from six to four cylinders, and adding an advanced fuel injection system has a different net value when done all together than when those technologies are considered separately and then simply added together. The car companies understand this. EPA's OMEGA understands this. NHTSA's VOLPE model does not. Instead, it considers each technology separately, not as a whole system.

As a result, OMEGA estimates very accurate cost results and fuel economy for combinations of technology five and ten years out. VOLPE is somewhat accurate at predicting costs a year or two into the future, but increasingly off base thereafter. This is a big problem, since the greenhouse gas rule has to project changes in fuel economy seven years out.

To make matters worse, NHTSA gets lots of its input information directly from the manufacturers, a group that is notoriously unreliable for providing accurate numbers to regulators. During a meeting at the Ann Arbor lab, Fiat CEO Sergio Marchionne once joked to me that: "We keep two books. One for us, and one for the government. Mrs. Oge, we will share with you the same book we have for us." The bottom line is that we are projecting the average total cost increase per vehicle in 2017 will be around $1,000. NHTSA's projection is closer to $4,000. If we can't get NHTSA's numbers to come down, we're going to have a hard time selling our proposal to White House economists.

★   ★   ★

As March continues, we have a series of weekly meetings in the spacious grandeur of the Old Executive Office Building. Located next to the White House, the building was built in what is described as "French Second Empire style." Mark Twain called it the ugliest building in America, but it has grand hallways and large rooms and its own storied history. The Old Executive Office Building is also where most people who "work at the White House" actually have their offices.

Positioned around a large conference table are Jody, Lisa Heinzerling, Ron Medford, me, the NHTSA team, and a number of White House economists from the auto task force. Chet and Bill fly out from Ann Arbor for the meetings and then back up to Michigan to work in the lab until late.

We are asked a lot of questions about our single national green-house gas program. How much greenhouse gas reduction will be achieved by the national program versus California's state greenhouse program? How can we give auto manufacturers a car that can be sold everywhere? When I say that we can get California to go along with our program, a critical point for the auto companies, the economists' response is blunt: "Glad to hear about California, but who cares if you and your sister agency NHTSA are so far apart on cost?"

Their skepticism is frustrating. Even in an environmentally-friendly administration, the gut instinct of many people is that EPA wants to put the car companies out of business—or at least we don't care if they succeed. Behind other questions is a deep but unfounded doubt about our science and numbers. At least our arguments will be battle-tested if we get sued.

NHTSA's deputy administrator, Ron Medford, tries to keep the working relationship positive. His standard response is, "We are work-ing to resolve our differences." But the members of the task force all know that our two agencies have very different implementation costs. NHTSA is projecting the single national program will cost thou-sands of dollars more per car than our estimates—a deal breaker. Since NHTSA is part of the Department of Transportation, many people assume they would have an edge over an environmental agency on questions about automobiles. So by mid-March, our meetings with the auto task force representatives have a definite theme: How do we know what we know?

At one meeting, one of the economists with a sharp face asks us, "How do you know the per-car cost in 2016 will be $1,000?"

Chet looks back and explains how his staff in our lab, working with industry experts, took apart a turbocharged car, reducing it into thousands of pieces, and then costed out every component. It is a well-established automotive industry discipline called "reverse-engineering."

Another economist follows up: "How would you know that fuel economy is going to be improved by four or five percent—are com-panies going to be using all hybrids and electric cars?"

We know the companies' five-year plans. We've worked with Ricardo, engineering consultants who have designed diesel engines in London, modified aircraft to fly without refueling around the world, engineered the gearbox used in the car that holds the land speed record, and collaborated with domestic and foreign auto companies. They are experts at putting together and integrating existing technologies. We make clear that all standards can be met with existing improvements to gasoline internal combustion engines and with combinations of proven cost-effective techniques as simple as better tires and more aerodynamic structures. Our rules aren't going to force GM to sell a million electric vehicles in two years.

Another economist skeptically asks how we know the indirect cost to the auto companies. In other words, even if the technology and labor only costs the $1,000 we are quoting, how have we included indirect costs like marketing, worker benefits, and maintenance?

"In 2007," I say, "we did a peer-reviewed study, and the National Academy of Science has agreed with our approach. It allows us to calculate indirect costs—nobody had done this work before."

★   ★   ★

During these meetings with NHTSA and the White House representatives, Jody plays the role of neutral facilitator between two agencies who don't agree each other. Between the meetings, she is constantly emailing Lisa and me with technical questions.

On Friday, March 6, 2009, Lisa forwards a multipage request from the two groups of economists responsible for advising the White House along with Jody's note: "OECC and NEC really need these answers by the end of weekend if there is any way for you to get them." Lisa prefaces Jody's message with, "She is asking for a tremendous amount of new information—by the end of the weekend. I don't see how we can get it done."

Lisa has canceled her trip to a climate conference in Italy with Lisa Jackson so she can be in town if our position needs to be defended.

Lisa's constant optimism and respect for our work goes a long way not just with me but with my whole team. She is constantly keeping the administrator informed of our progress, while offering guidance and support to the EPA team. The staff in Ann Arbor and DC are complaining to me constantly about the volume of work and the short turnaround times. But they remain driven by a sense that we may be on the cusp of a breakthrough. By mid-March, Chet and his team are accustomed to Jody's multipage Friday evening homework assignments.[1]

That doesn't mean they are happy. Chet has assigned Bill Charmley to lead the team assigned to this project. After working sixty-hour weeks, Bill spends the weekends with his wife and kids until around 8:00 at night. Then at 9:00 p.m. he has to slip off to the conference calls I have with him and Chet to go over our strategy for addressing the White House's questions. Not only is the workload tremendous, but for anyone at EPA who had spent 2007 and 2008 developing expertise on greenhouse gases, it is simply frustrating: "We did the work. We are experts. Why won't Jody and the economists listen to us?"

I tell them that those people don't understand the technical and cost issues, and this is our opportunity to educate them.

Typically, after the calls, Bill disappears into his office, sits behind his computer, and punches numbers into the OMEGA model until the early morning. At 2:00 a.m. Bill picks up the phone and calls another engineer, Rick Rykowski, who's slumped in front of a computer in his home office.[2]

Peering at a screen in his house a few miles away, Chet is equally miserable. Now he might as well stay up another couple hours before leaving for the airport and another flight to DC. Since February he has earned gold status on Delta. At 6:30 a.m., he leans back into his upgraded first class seat wondering if NHTSA is actively sabotaging the rules. Would they go that far to defend their right to set fuel economy?

"If EPA were stepping on my territory, I guess I'd try to kill them too," Chet thinks, adjusting his head rest. He wakes up as the plane touches down at Washington National, grabs his constant travel companion, a well-worn black Tumi bag, and hops a cab to the Old Executive Office Building adjacent to the White House.

Lisa Jackson tells me that Carol Browner is an advocate for our plan within the White House, but she is primarily occupied with the heavy lifting on the cap and trade legislation the administration has prioritized. The president's public words support strong increases in fuel economy, fighting climate change, and promoting energy security, but he has likely not yet been briefed on our proposal for the single national program to regulate greenhouse gases. Jody is immersed in the day-to-day facilitating of meetings between all the players. Everywhere she turns are pools of economists—the Council of Economic Advisors, the Auto Task Force, Vice President Biden's chief economist Jared Bernstein, and the Office of Information and Regulatory Affairs at OMB. Carol and Jody need anything we can give them to help defend this rule. So Jody has no second thoughts about calling Lisa or me at all sorts of odd hours, along with her established habit of sending long lists of questions with insane timelines.

★   ★   ★

On the morning of Wednesday, March 25, 2009, we are prepping for another auto task force subgroup meeting later that day. We have sent NHTSA our presentation, but they aren't happy. My counterpart at NHTSA, Ron Medford, is concerned that some of the slides we've created have too much detail for our non-expert audience. He's also concerned that the slides don't show EPA and NHTSA working together. He doesn't want a situation in which either agency is claiming to have the best answer. Or, as he puts it, where "one agency wins."

The problem is that we do have two different answers. And ours is correct. I email Ron back at 9:00 a.m. "I am very concerned that we can't show what we claim as significant progress given the huge disparity of the numbers. Can we talk at 12?"

At 9:17 a.m. Ron gets back to me. He likes the first few slides and won't argue a few small points.

9:24 a.m.: "Ron. I am in meetings until 12. Chet and Bill are flying this morning. Which are the detailed slides that you are suggesting to eliminate?"

9:30 a.m.: "Actually Peter shows these in his comments. For me Slide 7 has too much detail; Slides 8, 9 are repeats in more detail on what is said on earlier slides and Slide 13 is not needed."

For the rest of the morning, Ron and I negotiate over which slides can be eliminated from the presentation.

At 1:09 p.m. I email Lisa. There are signs that NHTSA is weakening: "Know that they took out their cost number for scenario 1 it will interesting [sic] to hear their explanation. For scenario 2 we are close." Then I go outside and get a taxi to the Old Executive Office Building on the far side of the White House.

At the building, I produce identification that corresponds to a pre-cleared name, date of birth, and Social Security number. Then I empty my pockets, walk through a metal detector, and continue down a hallway with ornate moulding running below fifteen-foot ceilings. Many presidents and vice presidents have used spaces in the Old Executive Office Building as their private offices away from the White House. In today's meeting place, Room 180, President Nixon once sat behind a wooden desk rigged with a microphone and recorded his meetings. The EPA creator's desk has long since been removed.

About twenty people are on hand for today's meeting. I explain exactly how we believed our plan would affect GM and Chrysler, the two companies in the worst financial shape.

Throughout the meeting, we stick to the line: our costs for domestic companies are reasonable, the figures match those included in the restructuring plans that the companies themselves prepared, the rules would affect GM, Ford, and Chrysler pretty much in the same way they affected the German and Japanese companies.

"There's a good chance we can get all the companies to agree to this program," I tell the task force.

Walking away from the building after the meeting, I smile at Lisa Heinzerling. Behind her, on the other side of its wrought iron fence, the White House shines. "An inch closer to convincing them," she says.

"I'll believe it when I see it," says Chet. He has a hard time believing anything is that easy in DC.

★ ★ ★

The next day, March 26, 2009, Steve Rattner, the lead advisor to the auto task force and his team meet with President Obama in the Roosevelt Room.[3] The fate of Chrysler and GM is put to a vote. Some in the room want to cut Chrysler loose; the others want to continue supporting the nation's third largest automaker. The president is the tiebreaker. He decides to facilitate Chrysler's sale to Italian carmaker FIAT, fire GM CEO Rick Wagner and clean out much of his board, and send more bailout money to GM.

As March winds into April, my staff and I spend our days—and many nights—answering Jody and the auto task force's many requests. Chet and his technical team are still negotiating with their counterparts at NHTSA, who are stubbornly clinging to their cost projections. John Hannon meets with NHTSA's lawyers to try and convince them that there is no legal impediment to our plan. Progress to narrow the gap between agencies is moving at a glacial pace. I hear that some of the NHTSA staff are also complaining about horrible working conditions and being overwhelmed by EPA. Nobody is happy. Something has to give.

# 12

# JUST GET IT DONE

"The rule will happen," says Jody. It is April 21, 2009, and Jody is looking at Beth Lowery, General Motors' vice president for environment, energy and safety. Beth, an auto industry veteran, has come to a room in the White House's West Wing to meet with Jody, Lisa Heinzerling, Chet, Ron Medford, and me.

Working with Chet, Ron, and John Hannon, I have already created a proposal that has the support of Lisa Jackson, and Ron has the approval of Transportation Secretary Ray LaHood. We've already looked at all the technology, the legal aspects and the net cost benefits to the consumer and completed an in-depth analysis of each auto companies' future product plans. The company-by-company evaluation provided insights to what each automaker could achieve over time.

After decades of inaction, this is historic because it is the first action the US government is taking to reduce greenhouse gas emissions from any source under the Clean Air Act. It calls for a 26 percent reduction in greenhouse gases from 2009 levels for cars and light trucks, which also results in the largest ever increase in fuel efficiency since the mid '70s. By 2016 the program requires new cars to average 35.5 miles per gallon fleetwide, which is equivalent to 250 grams per mile of carbon dioxide.

This plan has been developed with the input of Mary Nichols and Tom Cackette at California Air Resources Board and has their support. Jody knows she has the White House's backing. She's well aware that Beth's employer is on federal life support. But the final shape of any agreement is uncertain. If we don't get commitments from the automakers, some auto companies may file lawsuits in hopes of delaying the rule. But in this conference room, the power dynamics are clear. We have the better hand; we just need to play it correctly.

We don't get much out of Beth at our first meeting, but we agree to have her technical staff make the trip over to Ann Arbor tomorrow. Our engineers and GM's engineers have good relationships with each other and can talk the same language. The next day the General Motors technical staff visits Ann Arbor. At 5:55 p.m. Jody emails: "Beth Lowry called to say they (GM) had a good meeting with EPA folks today in Ann Arbor. Sounding good with GM. Call if you want. Otherwise, I might go get a beer."

Early on, Jody wanted to bring the Auto Alliance industry group into our negotiations. I worried this might screw up our negotiations before they even got started.

"Jody, we need to talk with individual companies," I said. "That's the only way to successfully negotiate." My conviction had only been reinforced during my team's negotiations over the 2000 diesel program, which substantially reduced the amount of deadly soot emitted by trucks and buses, as well as the 1999 Tier 2 car standards, which reduced the nitrogen oxide and hydrocarbons emissions from tailpipes and reduced the sulfur levels in the fuels. Individual companies have different levels of comfort with new regulation, and all want something specific in return. This technique gives productive negotiations a chance. In contrast, the industry groups just demand the lowest common denominator solutions.

GM is the first of thirteen car companies we are talking to individually to find out what they want and what they can give while not divulging what their competitors are requesting. We offer the companies confidentiality on their requests because it is the only way to maintain trust.

It turns out that GM's biggest ask is to get credits for producing flex fuel E85 vehicles, models that can run on 85 percent ethanol. Starting in 1993, Congress allowed car companies to receive credits towards their fleet CAFE requirements for every E85 car they build. The problem is that the vehicles don't compel drivers to buy the ethanol-rich blend; they still run fine on conventional gas. Compounding the problem is a lack of E85 availability at gas stations. As a result, only about 1 percent of the flex fuel vehicles actually run on 85 percent ethanol. The real-world carbon dioxide savings are slightly above zero. Recognizing this loophole, California has already eliminated credits for flex fuel. What's more, E85 is not groundbreaking—GM has already spent a lot of money developing and manufacturing these vehicles. Nonetheless, they claim they can't meet our standards without that credit.

Chet calls Tom out in California to see about including the credits in the agreement. Tom hates the idea, so I call Mary Nichols. I want Tom to take a look at what California's concession on ethanol flex fuel would mean in terms of total carbon dioxide reduction for a national program. Chet and his team run the numbers. It is only a small loss. Eventually Tom agrees to allow some flex fuel credits in the early years. GM is in.

The next day we agree to hold off meeting with the German companies—BMW, Mercedes, and VW—until next week. We're afraid they will react negatively because they are starting from a higher greenhouse gas basis and therefore will have to do more to meet the single national program's standards. Instead, we move forward with the domestics, Japanese, and other foreign companies while keeping a lower-level dialogue open with the German companies.

It's a good idea, but Ron and I are still shuttling back and forth between the other car companies. Every single one of them is trying to make their own agreement. Chrysler wants to protect its profit center, the Jeep Cherokee and other SUVs. Toyota wants credits for the Prius. The Prius is a great car that exceeds our carbon dioxide reduction and mileage goals, but it is already established in the marketplace. We no longer consider hybrids to be cutting-edge technology. Then there are specialty companies like Ferrari that have only a few models, none of which will ever get close to the proposed 35.5 miles per gallon target by 2016.

An agreement is starting to look good, and we are gaining momentum, though. Ford actually seems to be giving us real information about their future technologies. Chrysler is the least financially viable company, has the fewest forward-looking technologies in the pipeline, and is in the middle of a sell-off and corporate restructuring. Not surprisingly, they seem a bit lost and are unhelpful in discussions. The Japanese companies already have better fleetwide fuel efficiency and don't make as many demands on the process. The German companies are also on board.

Yet, while EPA and NHTSA are closer, we're still not on the same page on the cost benefits of the national program. Without unity, the president cannot announce the proposed regulation. We need help from the White House.

★   ★   ★

On May 1, 2009, Rahm Emmanuel walks into a high-ceilinged White House conference room without a jacket— a short man with a brisk presence and a quick mouth. Before anybody says anything, he is introducing himself around the table. There are ten of us, but he fixes on me.

"Where are you from?"

"Boston. Can't you tell from my accent?"

He laughs. "Don't bullshit a bullshitter."

"OK. I'm from Greece, but I lived in the Boston area when I first arrived here."

"I should have known." He quickly refocuses on the whole group. My counterpart from NHTSA, Ron Medford, sits on my left, my technical director Chet to my right. Rahm sits down at the oval conference table opposite Ron.

"Hey. Thanks for coming, you guys. Listen, this is a big issue for the president. There are different people in different agencies. You've got the EPA and the DOT. There are histories between all of you." Rahm looks at Ron and me. "But I want you to know that you have to get it done."

Transportation Secretary Ray LaHood starts agreeing and promising to do everything possible to get it together. "Everyone is working well together," he says.

This is not actually true, but no one is going to disagree right now.

Rahm seems used to people agreeing. "There is too much at stake here to let emotions and bureaucratic arguments get in the way of working together."

Lisa Jackson, the EPA administrator, and Lisa Heinzerling walk into the room. They were stuck in traffic.

"Hello," says Rahm, as they introduce themselves and sit down. "If we have to keep you locked up in a room, so be it."

Then he whispers to Lisa Jackson, "Just get it done."

★   ★   ★

On May 8, 2009, I walk through the various level of security outside the Old Executive Office Building and turn to walk through the heavy dark wood doors into what is called the War Room. During World War II, President Roosevelt met his advisors in this room. Sitting over the fireplace mantel is an oil portrait of George Washington in his general's uniform. Around the fireplace are plush leather sofas and reading lamps. I am headed toward yet another a conference table, trying to find resolution with NHTSA.

But after Rahm's intervention, this meeting is different. In the hallway as I am entering, Ron Medford pulls me aside. "We've made some corrections to our model. Our costs numbers are pretty close to yours—but not quite there yet."

At the table, everybody stays calm and professional despite the frustration and anger just under the surface. As Jody starts the meeting,

Bill Charmley is actually beaming. The daily pressure for data hasn't been much fun, but he is excited to finally have a chance to use all of his compiled knowledge. Under Bush, no one was listening; now a room full of people is pumping him for more.

NHTSA passes out sheets of paper with PowerPoint slides. For some reason, no one in government actually ever gives a full presentation. Chet quickly flips through their analysis. The NHTSA cost estimates are now much closer to the EPA estimates, but Chet catches a big issue. "It looks like you are assuming that several companies will pay fines for violating the fuel economy standards."

Under a single, unified national program for both greenhouse emissions and fuel economy standards, the penalty for noncompliance stands to change significantly. The CAFE standards allow an automaker to pay a "gas guzzler" penalty in the hundreds of dollars for each car that doesn't meet fuel economy standards. For some car companies—especially those that sold lower volume, luxury models—it was often cheaper to pay the penalty than meet the CAFE fuel economy standards. Manufacturers like Mercedes didn't think twice about paying the fee on a $50,000 car that was well below the fuel efficiency requirements. By contrast, under the Clean Air Act, noncompliance penalties are much stiffer. The law legally prohibits an automaker from introducing a new car into the market that does not meet emission requirements and imposes penalties as high as $37,500 per vehicle per day. Not surprisingly, automakers always choose to meet the Clean Air Act requirements. What Chet has noticed is that, under the single national program, violations of greenhouse gas emissions standards would trigger the Clean Air Act penalties, and the automakers' lenient "gas guzzler" penalty under the CAFE rules would be effectively voided.

This is a sticking point, but we are suddenly within distance of each other. After the meeting Ron and I talk.

"Can you get the EPA tech team to look at the NHTSA model?" says Ron. "And let them know what is wrong with it and how correct the cost numbers."

"Yes." The War Room is the scene of our last face-to-face battle with NHTSA on cost. Soon after that, the NHTSA and EPA teams are working together to reduce the initial $4,000 cost per vehicle to meet the standards down to EPA's estimate of approximately $1,000.

We are closing in on the biggest increase in fuel efficiency in nearly forty years.

★   ★   ★

As the date of the announcement draws near, President Obama's climate change czar, Carol Browner, decides to hold a meeting with White House and Office of Management and Budget economists. Typically, any rulemaking would have had to go through a much longer process to get to this point. This is only an announcement of an agreement that has to go through the formal regulatory process before it becomes a legal requirement.

I stand in a room in the Old Executive Office Building, explaining that the new technology in our program would boost the average cost of a car by about $1,000 but net consumers about $3,000 savings in fuel costs. Michael Greenstone, Chief Economist at the White House Council of Economic Advisers, gestures at one of the slides EPA has handed out. "How did you come to the conclusion that boosting fuel economy in cars and trucks would benefit consumers by $3,000?"

Greenstone has close-cropped dark hair, jarring blue eyes and specializes in evaluating the cost benefit analysis of environmental regulations. He has been grilling me for the past ten minutes.

Seated next to me, Chet explains that the analysis includes both fuel savings and the benefit of carbon reductions. Of the $240 billion in benefits, 80 percent comes simply because cars would use less gas.

There is a pause in the room while Greenstone checks his multiple BlackBerries.

"The consumer won't fully value these fuel economy benefits, so we should discount them by 50 to 80 percent."

Greenstone's comment nods toward a fundamental philosophical problem in our dealings the White House economists. Even if they believed all our numbers and didn't think our proposal would unduly burden a flailing auto industry, academics like Greenstone would still worry that we are messing with the magic of the market.

Consumers already have the option to buy fuel-efficient cars, he says, and they often chose not to. In Greenstone's world, if consumers truly value fuel savings, then they would have been buying fuel-efficient cars all along. If the benefits are so clear, then we don't need regulation. The car companies will make these cars because the consumer demanded them. Instead, consumers' behavior suggests they do not value fuel economy. That behavior needs to be included in the cost-benefit equation.

I breathe before turning to Greenstone. "Well, I'm just an engineer, so I can't compete with you on economic theory. But if I save one dollar at the pump, I'm not going to throw fifty cents into the wastebasket."

He looks at me, then down at his BlackBerries again, wordlessly dismissive.

The idea that the market functions perfectly is a powerful political and theoretical obstacle to fuel economy regulations. Carol tables the issue for the moment by suggesting that we have follow-up discussions on this issue during the interagency process. That is to say, after the agreement is announced. Lisa Heinzeling will have to personally work with the powerful Office of Management and Budget Director, Cass Sunstein, later on. For now, the economists can't do anything but ask questions.

★   ★   ★

Getting closer to an agreement doesn't make our lives easier. Jody continues to be under intense pressure about the cost numbers for the national program. Despite my advice, she gets nervous about our cost projections. She calls me and says she is going to contact the auto

companies to confirm our cost projection of $1,000 in average vehi-
cle cost under the program. I grit my teeth together.

"Jody, you know what they will say. They always estimate that any
regulation will cost far more than it actually does. But, I look forward
to being pleasantly surprised."

This is how the industry operated for the entire forty years' existence of the EPA and before. They fought new mandates on safety
and health—from seatbelts to catalytic converters, from air bags
to anti-lock breaks—by claiming the cost would put them out of
business.

"I understand, but I still need to hear from them," she says and
hangs up. But when the numbers came back from GM and Ford, they
are roughly $1,000 a car. This never happens. I am indeed pleasantly
surprised. Maybe their near bankruptcies have scared them straight.
For the moment.

★   ★   ★

As the deadline approaches for confirming the details of what is now
known as the 2012–2016 Obama National Policy for Car and Light
Trucks, we are still in tight negotiations. The president's announcement of the plan is already scheduled for Tuesday, May 19.

On Thursday, May 14, our schedule looks like this:

12:00 Daimler
12:30 Brad Markell of the United Auto Workers
1:00 Nissan
1:30 BMW
1:45 VW

As the agreement comes together, we are keeping the environmental groups and the states informed. During one call, the National
Resources Defense Council's Roland Hwang is worried that we're

making too many trade-offs. "Trust me," I say. "It will be good." My word is the best I can do for now.

Simultaneously, Mary Nichols in California is feeding updates to the other states that are part of their coalition. With the agreement's announcement imminent, Jody needs to place courtesy calls to the states that have been supporting regulating greenhouse gas emissions for years. It's a perfunctory detail that she crams between her more serious unfinished business, until one assistant state attorney general decides to press his luck.[1]

"We have some changes we'd like to see," he says.

Jody pauses for two seconds. Over the past five months, she has ushered this agreement past grinding battles with NHTSA and a horde of skeptical White House economists, and she has gained the acceptance of major car companies. This is not going to be another negotiation.

"The president of the United States is getting ready to announce this agreement! No changes!"

Then she hangs up and begins inviting governors to the event.

On Friday, the 15th of May, Jody sends out letters of commitment for the car companies and California to sign. My lawyer, John Hannon, managed to draft the unorthodox binding document that isn't a rule and doesn't even have the force of law. Once the president announces the agreement, it will, however, carry the weight of public opinion. The car companies and California are to sign and return them before the ceremony at the White House on Tuesday.

★   ★   ★

The next day, I take the train up to New York. My daughter Nicole's bridal shower is that afternoon at a friend's apartment. I had put together a slide show of old photos of Nicole. I joke with her friends about some of Nicole's childhood stories. I can tell I'm making her a little nervous. She has a "what-is-mommy-going-to-say-next?" look. My phone rings, and I glance down. Jody. My heart sinks.

"Excuse me," I say to my daughter's friends and make my way out onto the balcony of the loft.

"Ford has a problem with the agreement. Sue is threatening to pull out."

Sue Cischke, Ford's lead negotiator, has told Jody that the final agreement doesn't include the credits we had promised Ford. The credits would let them offset later requirements by selling more clean cars early in the cycle.

"She's bluffing," I say. The credits weren't worth nearly enough to blow up the agreement. I figure they wanted something else.

As my daughter's bridal shower continues inside, I find myself in a conference call with Sue Cischke, Lisa Heinzerling, and Ron Medford, who is calling in from his weekend home.

Jody is similarly flustered. Before calling me, she had been about to eat dinner with a friend. But just as her prime steak and wine arrived, her BlackBerry buzzed. She looked down at the food, apologized, and walked away. Lisa Heinzerling was at dinner at a neighbor's house and also had to abruptly leave to join the call.[2]

Eventually, I convince Sue that her credits are there. The crisis is averted, and I can go back inside and Ron can get back to his weekend off. Jody's steak and wine will have to wait for another day.

★  ★  ★

May 19 is a beautiful spring day. There is a long line of invitees being checked by the security guard outside the West Wing of the White House. Inside, a table with water and lemonade is set up on the very green lawn. There is also an amusing collection of people.

It is no surprise to see Jody and members of the EPA team, although Chet couldn't convince Bill Charmley to come down. Bill asked to spend the day with his family, a too rare occurrence in recent months. Mingling with us are some White House lawyers and people from the Office of Management and Budget. Across the grass I see Ron Medford and the NHTSA team. Mary Nichols had to force Tom

Cackette onto the plane for the flight from California. Tom was sick of the process by the end and wanted no part of Washington, but he seems to be enjoying himself now.

Governor Schwarzenegger is proudly making his way around the crowd in cowboy boots. He is a car aficionado and talks with some of the thirteen auto CEOs waiting to get on stage with the president. There is also a large contingent of people from the major environmental groups, including Michelle Robinson, director of the Car Program from the Union of Concerned Scientists. California representative Fran Pavley is here to witness the national version of her landmark greenhouse gas law announced. I talk with Carol Browner and her deputy, Heather Zichal. To their right stands John Dingell, the powerful Michigan Democrat who had foiled fuel efficiency increases in the House of Representatives for decades.

The crowd is somewhat surreal. We are surrounded by our colleagues, allies, and supporters as well as people who spent decades challenging mileage increases and greenhouse gas reduction requirements. But, this being Washington, almost everyone is taking credit for the historic agreement.

Finally, the president takes the podium flanked by the CEOs, union representatives, Transportation Secretary Ray LaHood, and our administrator, Lisa Jackson. It is almost exactly four months since President Obama was sworn in before record crowds celebrating the coming of a brighter future.

"In the past," he says, "an agreement such as this would have been considered impossible.

"That is why this announcement is so important, for it represents not only a change in policy in Washington, but it is the harbinger of a change in the way business is done in Washington. As a result of this agreement, we will save 1.8 billion barrels of oil over the lifetime of the vehicles sold in the next five years. And at a time of historic crisis in our auto industry, this rule provides the clear certainty that will allow these companies to plan for a future in which they are building the cars of the twenty-first century."[3]

Just ten months earlier EPA administrator Steve Johnson publicly insulted our proposed greenhouse gas rule in the official register of the federal government, throwing years of work into the scrapheap. Now we have turned that work into a five-year program that will reduce greenhouse gases by about 960 million metric tons, or the equivalent of taking 177 million cars off the road or shutting down 194 coal-fired power plants. For the first time in history, we have set in motion a national policy aimed at both increasing gas mileage and decreasing greenhouse gas pollution for all new trucks and cars sold in the United States of America.

As the president walks off the stage, John Hannon yells, "Thank you!"

A reporter calls out, "What about after 2016?"

Secretary LaHood smiles. "Let's let everybody get some sleep first."

# 13

# 722 JACKSON PLACE

For all the celebration, President Obama is not announcing a new regulation but a letter of agreement. Turning that document into something enforceable is a longer bureaucratic process that takes up much of the next year. We first release a proposal and then request public comment, present it around the country, seek the opinions of stakeholders, finalize the proposal, and respond to all comments. After our four-month effort to put together the agreement, we don't release the final rule until nearly eleven months after its announcement, April 1, 2010. I don't know who picked April Fool's Day. To give the car companies sufficient time to invest and make the needed technology changes, the rules don't go into effect until the 2012 model year.

The official title of the national program is now the 2012–2016 Light-Duty Vehicle CAFE and Greenhouse Gas Standards. It requires an automaker's fleet of 2016 vehicles to a fleet-wide average of 250 grams per mile of carbon dioxide in tailpipe emissions, an amount that is equivalent to 35.5 mpg. This figure includes reduction in both carbon dioxide and the chlorofluorocarbons used in air conditioning systems.

But, as groundbreaking as the new standards are, a bureaucratic gap that I mentioned earlier renders the accomplishment somewhat confusing for consumers. At issue are the EPA window sticker fuel economy values, which reflect real world performance, and the CAFE values based on outdated testing procedures that can only be changed by Congress. As a result, a 2016 car that is rated at a CAFE value of 35.5 mpg will have an EPA window-sticker that reads 27 mpg, about 20 percent lower.

In 2016, not every vehicle will have to meet the CAFE 35.5 mpg requirement, but rather the fleet of vehicles sold in the United States will have to average 35.5 mpg. This allows significant flexibility for an automaker. For example, GM can continue to offer their large SUVs as long as they meet the standards by also selling much more efficient cars, like the electric Chevy Volt.

Trucks will continue to have lower requirements than cars. Because they are larger vehicles, trucks take more energy to run. As a result, the same technological improvements in a truck will cost more than in a car.

Also the improvements for each vehicle will depend to what is called the "footprint." Footprints are determined by the distance between the track width—the space between axles—and the wheelbase, or distance from the front to rear axles. As a result, for example, a compact car like a Honda Fit will have to get 41 mpg while a Chevy Silverado truck will need to get 24.7 mpg.

The compliance costs for the industry are estimated at less than $52 billion, or an average cost per vehicle of $950 in 2016. For consumers who buy a more expensive vehicle but save money on gas, this translates to a buyback period of three years, after which the more efficient vehicles are essentially putting money back into their pocket. Over the average life of a vehicle, Americans will come out with about $4,000 in total savings on fuel.

The net result is a program that will save Americans 1.8 billion barrels of oil and provide total benefits of $240 billion. It will reduce carbon dioxide emissions by 960 million metric tons of greenhouse gas. By 2030 overall vehicle greenhouse gas emissions will be reduced by 21 percent.

There was also a second major element of the 2012–2016 rule that was not only historic but had implications far beyond the transportation sector. As part of the process, the EPA finally released its greenhouse gas endangerment finding in response to the 2007 Supreme Court decision. Our analysis was very similar to the 4,000-page finding that we had sent over to the Office of Management and Budget in December of 2007 only to have them refuse to open it.

In the finding, EPA determined that the six greenhouse gases in the atmosphere—carbon dioxide, methane, nitrous oxide, hydrofluorocarbons, perfluorocarbons, and sulfur hexafluoride—are indeed a danger to the public health and the welfare of current and future generations. To support this rule, EPA declared that greenhouse gas emissions from new motor vehicles and their engines, namely carbon dioxide, methane, nitrous oxide, and hydrofluorocarbons, cause and contribute to climate warming. After stonewalling by the Bush administration, this time our science became a rule. Not only that, it provided a precedent for regulating the greenhouse gases emitted not just by vehicles but by power plants and other stationary sources of climate change–inducing pollution. Congressional refusal to pass a climate bill can't stop the Obama administration from using the Clean Air Act to regulate carbon pollution from utilities.

★ ★ ★

It doesn't take our colleagues in California long to shake things up. Within a few weeks of the rule's release, the state floats a white paper laying out its plans for greenhouse gas regulation through 2025. Their proposal includes more stringent standards on greenhouse gas emissions and requirements for zero emission vehicles. Now the car companies are nervous all over again. To provide stability in the marketplace and avoid producing two cars for the US market, they had just agreed to set nationwide standards through 2016. California's aggressive new proposal reopens the possibility of splintered markets and unknown new regulations. The automakers talk to Congress, Congress talks to the White House, and on May 21, 2010, the president orders EPA and NHTSA to work with California to develop new standards from 2017–2025. We're off again.

A few days later, Chet, Jody, and I are in a taxicab in downtown DC headed over to the Justice Department. "This is an amazing opportunity," I say. "You need to take this on. It's going to be bigger than the first agreement."

Jody seems a little distant. She has more or less decided to head back to her faculty position at Harvard, and before too long she is back up in Cambridge.

After all the battles we've fought together, I am sorry to lose Lisa, but we've got a lot of momentum and I'm not going to shy away from another historic goal. Before we can propose any rule, however, we need to know what is technically possible for passenger vehicles fifteen years from now. This is unknown territory.

The first agreement didn't require us projecting more than seven years out, relied entirely on already existing technology, and was based on the California greenhouse gas car program. Very early in 2009, we had settled on an endpoint of 250 grams per mile carbon dioxide and a fuel economy goal of 35.5 miles per gallon. Now we're looking a decade and a half into the future and are operating with a lot more latitude, increasing our opportunities for both success and failure. In short, we need a lot more data before we can even consider proposing a rule.

Chet's engineers went back to work, produced a detailed evaluation of various automotive technologies, and modeled the potential future scenarios of ways the industry could meet more stringent GHG standards. The team evaluated over 30 technologies with potential to reduce GHG and improve fuel economy. This technical document was later put out for public comment by the fall of 2010.

In parallel, Tom Cackette from California's Air Resources Board, Ron Medford from NHTSA, and Gary Guzy from the White House's Council on Environmental Quality, Chet, Bill, and I were back on the road. From June to August, we met or had discussions with about seventy stakeholders, including car companies, suppliers, unions, academics, and environmental and consumer groups among others.

We visit Ford's headquarters—a modernist cube—and Chrysler's mall-like complex, both of which are located in the Detroit suburbs. We go to GM's towering new headquarters on the Detroit River overlooking Canada. We meet Hyundai and Honda in California. BMW, Mercedes, and VW come to us in Washington.

At each stop, we sit at a conference table and discuss what it would take to improve fuel economy by 3 to 6 percent annually for 2017–2025. Then we wait for the car companies to tell us what they can do.

None of the domestic companies will commit to a specific number, but Ford and GM both seem confident about their technology. Chrysler is reorganizing with their new owners at Fiat and still doesn't have a lot of exciting stuff in the pipeline. The companies' representatives talk about electric and hybrid cars, but not all of them were committed to putting electric vehicle technology on the showroom floor for a while.

The Asian companies are a different story. Honda sends us some useful advance information and flies a number of their senior people from Japan over to California for the meeting. Soon after introductions, I speak up. "It looks like you are saying that at minimum you guys could do four percent."

The executives don't answer immediately. They stall and discuss problems with that rate of improvement. But the company prides itself on its forward-looking fuel efficiency, and we eventually hear, "You are right. We could do four percent."

The last day of our California trip, we meet with Hyundai. They are even more enthusiastic. They support 6 percent, our maximum proposed fuel efficiency improvements. "At a minimum," says John Krafcik, the US Hyundai president, "you should do 6 percent."

Back in DC, BMW is the most useful from a technological perspective; Mercedes and VW aren't bringing much to the table. And no one will commit to a number. We leave these meetings with a better understanding about each company's technology portfolios but not much of an idea what improvements they could live with.

Even though we don't have everything we want, the president has ordered us to move things forward by releasing a technical report and then working toward what's called a Notice of Intent for a new rule. So, in early 2011, Chet, Bill, and Mary fly in for a meeting with Ron Medford, NHTSA administrator Dave Strickland, Gary Guzy, and me. Also present is Gina McCarthy, a tough, smart, and hardworking woman from Boston who is the EPA's new assistant administrator for air and radiation and my new boss. A few years previously, I had worked with Gina in an effort to reduce nitrogen oxide emissions from aircraft when she was with the Massachusetts Executive Office of Energy and Environmental Affairs. It didn't work out, but I did get to know her. Gina is down to earth, demanding, and a good listener. We always agree on important decisions. I like to tease her that we understand each other so well because both have a Boston accent.

We gather at the EPA. Since we've already established a timeline, the biggest unknown in this agreement is what the target for greenhouse gas reductions and annual rate of fuel economy improvement will be. We propose 5 percent, toward the high end of our suggested 3 to 6 percent we'd brought to the auto companies. The numbers make NHTSA's Ron Medford a little nervous, but California's Tom Cackette is quite happy. In fact, our proposal is a straw man. Not only are we sure the car companies won't accept it, but we never had any intention of pushing for 5 percent for light trucks outright. The Ann Arbor tech team doesn't think it's realistic that Chevy could improve the fuel economy on, say, a Silverado pickup by 5 percent year on year starting in 2017. But we are expecting pushback from the companies no matter what we bring to the table. This proposal isn't outrageous, and it gives us a little bit of wiggle room.

★　★　★

While we are discussing how much of a reduction in carbon dioxide emissions is reasonable, lobbyists and congressional staffers are holding their own meeting just a few blocks away. The new Republican Congress has an interest in stopping just about anything the president supports, and the EPA is always a good target for conservative attacks. Some of the Republicans are adamant about killing EPA's efforts to regulate greenhouse gases.

On March 15, 2011, the House Energy and Commerce committee votes to strip EPA of its Clean Air Act authority to "promulgate any regulation concerning, take action relating to, or take into consideration the emission of a greenhouse gas to address climate change." The bill also repeals several rules and actions, including EPA's scientific assessment that identifies greenhouse gases as pollutants.

The vote reeks of both narrow political interests and petty attacks on President Obama's signature environmental achivement. Our finding that greenhouse gases are dangerous pollutants was initially a response to a decision by a majority Republican-appointed Supreme Court and a subsequent exective order by President George W. Bush. The 1990 Clean Air Act was signed into law by Bush's father, also a Republican president. The only rule regulating greenhouse gases to date was agreed to by almost every major auto manufacturer in the American market and has faced no legal challenges from the industry. Despite being a breakthrough, our greenhouse gas rule is hardly a unilateral piece of regulation muscled though by an overreaching executive branch. But Congress continues to play its Janus-like role in protecting public health and the environment. Twenty-one years ago, it created a strong, updated 1990 Clean Air Act; now it's trying to shred our ability to fight global climate change.

★   ★   ★

Over the next few months, negotiations with the automakers continue without any major breakthroughs. By the end of June, the pressure from the White House to make progress is increasing. We have the precedent of the first rule as a blueprint, all the same technical expertise, a better working relationship with NHTSA, and a negotiating position with regards to numbers and timetables, but there is still something missing.

On a discussion with Gina, I say, "I am worried that without the White House involved, NHTSA staff won't go ahead with a proposal this year. They don't know it's a priority—and neither do the companies."

What we are missing is Jody, and, since she is staying put in Cambridge, Gina calls up Heather Zichal at the White House. Heather eventually gets Obama's new chief of staff, William Daley, to chat with Lisa Jackson, Gina McCarthy, and Transportation Secretary Ray LaHood. Daley makes it clear that NHTSA and EPA need to move forward quickly. To push forward negotiations, he brings in a man named Ron Bloom who served on the auto task force.

Ron Bloom has very short-cropped hair, an MBA from Harvard, and a reputation for knocking heads together. He has spent time both in the world of investment banking and as a representative for steelworkers and other unions. He is known for crafting creative solutions in complex negotiations. He already knows the CEOs of the Big Three and was impressed with our Ann Arbor lab during his visit in August 2009. Gina considers him a very powerful negotiator, but I am holding back my views on Ron Bloom for the time being. He now has the tough job of finalizing negotiations with the auto companies in time for the president to announce the agreement by late July.

For Ron Bloom, these negotiations are not only important for both climate and the nation's energy security but probably the only chance to get anything done on either for the foreseeable future. President Obama's climate bill had died in Congress even before the Republicans took over the House. He would only get an important, signature achievement on climate change through his regulatory agencies, specifically EPA.[1] Ron Bloom is now the administration's candidate to push forward negotiations in time for the president to announce the agreement by late July. The time was short and the midsummer Washington heat was on.

★   ★   ★

In early July 2011, I walk up the five steps of 722 Jackson Place, a nineteenth-century brownstone across the street from Washington, DC's, Lafayette Park and within hailing distance of the White House. Because of its prime location near the president, the houses on this block were some of the most sought-after private residences beginning in the early 1800s. But, as with most properties near the White House, the federal government bought them up during the twentieth century and placed executive agencies inside. This building belongs to the White House's Council on Environmental Quality. Today, CEQ is hosting our meeting with Ford.

Once inside, I turn left into a large conference room. On the right is a rectangular table. Soon, Ron Medford and his NHTSA team are seated at the far side. Gina, Chet, and I sit opposite them. Ron Bloom limps in, a cast on his recently broken foot from a running accident. He sits at the head of the table facing the door. To his right is Sue Cischke, a thirty-five-year veteran of the auto industry and the woman whose last-minute call caused me to miss part of my daughter's bridal shower two years ago. Of course, I don't hold any grudge. I don't have time to.

Soon after the meeting starts, Chet lays out our proposal for Ford to increase the fuel efficiency of its cars and trucks by 5 percent every year between 2017 and 2025. Sue's mouth flattens into a thin line and she rolls her eyes. I'm sitting next to the door, half expecting Sue to get up and walk out. Gina and I exchange glances. Ford is probably the most forward-looking American company. If it won't stay at the table, we'll definitely lose GM and Chrysler.

"You guys don't know what you're talking about," says Sue. She flashes to me. "Were you hearing what I was saying? We showed you everything we had. You have totally overestimated what we can do with trucks." During our discussions regarding the 2016 standards, she had told me that those standards would already be hard to reach for pickups. Trucks, especially the F150, are where Ford is making a lot of its money. She is talking to me now, but I'm still not sure she won't walk out.

"Sue," I say, "is it the cars or is it the trucks?"

She doesn't answer me directly, but she spends most of the shortened meeting talking about trucks. She isn't happy but does agree to have our tech teams get together in Michigan the next day. We have always assumed that our proposed truck improvements will be negotiated down, but if we can get them to share their data with us, then we can zero in on what they really can do. Our plan is to get Ford to agree and then bring in GM and Chrysler. If the domestic automakers are in, it will be hard for many of the foreign manufacturers to say no.

Out in Ann Arbor the next day, Bill Charmley begins talks with his counterparts at Ford. After years of working with us, Ford is more comfortable with the EPA and its lab than with any other federal agency. Ford begins sharing their technical data on sheets that they retrieve at the end of the meeting. "There's no way," an engineer says, "we'd email you or show you this stuff at Jackson Place—not with all those other people around."[2] For the next few weeks, Bill has his team of engineers and modelers looking at all the new information and trying to come up with technical solutions to the problems remaining.

Meanwhile, we are at the Council on Environmental Quality's office near the White House, holding multiple parallel meetings with car companies in the two downstairs conference rooms and in the borrowed offices or hallways that sit up a narrow staircase. The process is simultaneously ever changing and frustratingly slow.

Ford has invested more in understanding the government's technical framework. Now they are getting into the weeds, seeing the evolution of vehicles and mileage. We are beginning to create a productive technical dialogue.

GM doesn't have the same kind of engagement. Maybe their bankruptcy proceedings had sapped some energy from this—or maybe

they were just being the difficult old GM. Either way, they are claiming to have tried everything.

Chrysler continues to express concerns about the ability of their
trucks to make significant improvements beyond 2016. Toyota and
Honda both want a special break for hybrids, but we want this
agreement to encourage the growth of new, revolutionary technologies. Hybrids are great, but not as game-changing as fully electric
vehicles or fuel cells. The German companies want special breaks
for advanced clean diesel vehicles, but the diesel engine is hardly a
breakthrough. It has been around for a hundred years, and companies already know how to do it in a cost-effective way. VW simply
says they've run the numbers, and it doesn't work. Hyundai is still
ready to do 6 percent.

Car companies always feel that the regulators are being aggressive
and don't understand the technologies, costs, economic impacts, or
market. We are there to talk them down. For our part, we also assume
they aren't showing us all their cards—at least not up front.

The building hums with haggling and frustration and small breakthroughs all day and night. Trust is a valuable but fleeting currency.
Mary Nichols is having a hard time believing practically any of the
numbers companies are offering.[3] More often than not, she's probably
right not to. Meanwhile, NHTSA is having a hard time with some
of the legal issues. But we need the confidence of everyone in the
building, or we'll have a hard time coming up with any agreement.
Between sessions, some of us step onto to the sweltering patio out
back to grab a Coke or a piece of pizza. At night we huddle over cups
of coffee inside.

Ron Bloom isn't nearly as involved in the details as Jody was, but
he is still an extraordinary negotiator, moving the agreement forward.
Gina was right; he is great. In one late night session, he becomes
increasingly annoyed at the inflexibility of NHTSA's attorneys. When
we reach a creative solution for one problem at about 10:00 p.m.,
Ron checks with the lawyers in the room.

"Yeah, we can do this," says our team.

Ron automatically turns to NHTSA's team. "I suppose you can't."

"Right."

We go back to discussing other options. A few minutes later, the NHTSA attorney raises his hand. "Actually, I think we can do this."

Ron stares back. "Son of a bitch! Now we can do it? Where did that come from? What page was that on?" His sudden eruption leaves us laughing.

Eventually, we get the domestic companies interested with two proposals. One is a 5 percent annual increase in fuel efficiency for cars and a 3.5 percent increase for trucks. Because 2025 goes beyond the automakers' normal five-year-increment planning process, we offer a review to measure the car companies' success in 2018 and determine whether the required efficiencies should either stay the same or be lowered. But the agreement on trucks upsets Honda and Toyota, who accuse us of bias toward domestic companies. At one point Toyota's representatives threaten to stop production of trucks in America, while Honda's team is raving about adjustments they want for small cars. Meanwhile, Mercedes and VW want special consideration for diesel cars.

Then BMW moves toward signing on after we offer them incentives on electric car technology. Chrysler, because of the merger with Fiat, now has a natural gas car with a very small gas-powered engine that they feel good about.

But by the third week we still don't have a proposal locked down. Ron takes me aside to say that Ford is getting pushed by the Michigan congressional delegation not to make a deal. "If we don't get GM, then Ford will split." And if Ford isn't in, then Chrysler will drop out. Fortunately, the dominoes fall our way. On the third Sunday, Ford and GM come on board. Chrysler follows. Then Toyota and Honda come in, somewhat unhappily.

★　★　★

On July 29, 2011, a fleet of shiny new cars, SUVs, and pickup trucks serves as a gleaming backdrop for President Obama as he strides across a brightly lit convention center stage. He quickly shakes the hands of thirteen smiling auto CEOs and senior executives before stepping to the lectern and announcing some of the stiffest new regulations on automobiles in decades. The executives sit listening agreeably, even proudly, to the plan that would compel their companies—representing 90 percent of the American market—to double the fuel efficiency of their products and cut greenhouse gas emissions in half by 2025.

Thirty feet away, I sit in the front row listening as what had once seemed impossible was announced in the president's crisp, confident voice. Then, as the press and crowd empty out of the convention hall, White House staffers usher my EPA team to a bus and whisk us up Pennsylvania Avenue to the White House.

Inside, a still smiling but more exhausted-looking Obama greets us. He shows us into the Oval Office, the famous room where he does his day-to-day work. The past few weeks his job has mostly consisted of battling Congress to a standstill over the federal debt limit. An issue of *Sports Illustrated* peaks out from underneath a briefing folder on his desk.

As we are getting ready to take a picture with him, Obama turns to us and says, "How in the world did you guys do this? You need to come and help me with Congress on the budget discussions."

★　★　★

More than a year after Obama's Rose Garden announcement, we release what are formally known as the National Program for Model Year 2017–2025 CAFE and Greenhouse Gas Standards. The program has many of the same advantages as the 2012–2016 rule, but the benefits go much deeper. In 2025 a new car will cost $1,800 more, but the consumer will get up to $7,400 in fuel savings. It will save an additional four billion barrels of oil and two billion metric tons of greenhouse gas emissions.

The combined 2016/2025 program will save the consumers $1.7 trillion and the country 12 billion barrels of oil. It will reduce oil consumption by more than two million barrels daily, as much as half of the oil we import from OPEC each day. Even more important, the program will reduce greenhouse gas emissions by six billion metric tons over the life of the program. This is equivalent to the total amount of carbon dioxide emitted by the United States in 2010, or what the Amazon rainforest absorbs in three years.[4]

Picking up where our previous rules leave off, this regulation continues the 5 percent average rate of mileage improvement for cars through 2025. Trucks still move at a slower rate, improving fuel economy 3.5 percent through 2021 before moving up to 5 percent through 2025. In 2025, new cars and trucks will get a combined 54.5 mpg, or 163 grams per mile of carbon dioxide, about 40 mpg in the real world.

The new program contains incentives to encourage early adoption and introduction of advanced technologies like electric vehicles, plug-in hybrid electric vehicles, fuel cells vehicles, and the hybridization of large pickups. There are also incentives for natural gas vehicles and credits for technologies like solar panels that have the potential to achieve real-world greenhouse gas reductions.

★   ★   ★

On August 28, 2012, the same day that these new standards are proudly announced, Mitt Romney becomes the official Republican Party candidate for president. In his acceptance speech, Romney mocks President Obama's action on climate change. "President Obama promised to begin to slow the rise of the oceans and heal the planet," says Romney to audience laughter.

It is not surprising that the Romney campaign later criticizes the new greenhouse gas rules as "extreme," adding that the standards would limit consumers' choices when they shop for a new car. "The president tells voters that his regulations will save them thousands of dollars at the pump, but always forgets to mention that the savings will be wiped out by having to pay thousands of dollars more upfront for unproven technology that they may not even want," says Andrea Saul, Romney's spokeswoman.[5]

Continuing to ignore the challenge of global warming is no longer a viable option, but you wouldn't know it from the Romney campaign's rhetoric. They remind me that some of the same forces that derailed laws to raise fuel economy in the 1980s and made sure the Kyoto Protocol remained a meaningless piece of paper in the 1990s will be targeting our breakthrough regulations.

But, despite all the delay in finally acting to slow the buildup of greenhouse gases in the atmosphere, I believe we are finally moving into the future. First, Mitt Romney's rhetoric didn't win him the White House. Second, we are more likely to hear current car industry executives promoting cleaner, smarter vehicles than complaining about them. Just as important, the auto industry has been joined by a bipartisan coalition of financial and military leaders, politicians, scientists, economists, doctors, and progressive energy companies, among many others, in recognizing the necessity of confronting climate change.

These are baby steps in the right direction, just as my story of the first federal regulations limiting greenhouse gas is only a prelude. This breakthrough must be followed by a revolution. Starting immediately, we need to move much further and much faster to reduce greenhouse gas emissions. In the transportation sector we need cycles of innovation that outpace anything we've done before. We need rapid adoption of new, cleaner technologies. We need people around the world to reimagine personal transportation. But even though this challenge is perhaps the greatest we have ever faced as a species, I believe we can meet it—and that the cars of the future will have a central role. But this is a much bigger story that will play out globally as an ever changing tomorrow rushes forward to meet us.

# PART THREE

# IMAGINING TOMORROW

## 14

# THE "FOUR WHEELS" DRIVING
# THE CAR OF THE FUTURE

In December of 2009, just seven months after President Obama had announced the first US action to regulate greenhouse gas emissions, representatives from 119 countries met behind the rounded glass façade of Copenhagen's Bella Center. The event, a United Nations climate conference, brought together leaders of nations that account for at least 80 percent of the world's greenhouse gases, including Obama. By his presence, the president said he was signaling that the United States would now be taking a leading role in global climate discussions: "The time has come for us to get off the sidelines and shape the future that we seek; that is why I came to Copenhagen." Because his words came close on the heels of our breakthrough rules on cars and trucks, I hoped the United States might be on a roll.

Unfortunately, the meeting was not entirely successful. Spats between developing and developed nations erupted, including a blame game between China and the United States. The agreement that emerged, known as the Copenhagen Accord, was not legally binding for any of the participants. Both shortcomings were reminders of the myriad difficulties in unifying nations to combat climate change.

The meeting also at least hinted where the United States—and the rest of the world—needs to set its greenhouse gas reduction goals going forward. The nations at the Copenhagen Summit issued a "recognition" that scientists have determined the world needs to limit future temperature rises to below two degrees Celsius (3.6 Fahrenheit) above pre-industrial levels. Failure to do so would have devastating effects on the earth's ecosystems, public health, and the global economy. Using just one metric, a July 2014 White House report assessed that "a delay that results in warming of 3° Celsius [5.4 Fahrenheit] above preindustrial levels, instead of 2° [3.6 Fahrenheit], could increase economic damages by approximately 0.9 percent of global output."[1] Today, that would represent a $150 billion hit to the United States. Globally, the annual damage would be worth more than the combined total assets of ExxonMobil, GM, Chrysler, and Coca-Cola. And every decade of inaction would increase the cost by roughly 40 percent.

But when the countries were asked to set voluntary greenhouse gas reduction goals for 2020, the gap between scientific necessity and global action was huge. The United States, for example, committed to a 17 percent reduction from 2005 levels. China promised a 40–45 percent drop. But as the IPCC has repeatedly concluded, to keep future temperature rises below two degrees Celsius and prevent catastrophic climate change, globally we must reduce greenhouse gas emissions by 50 percent by 2030 and 80 percent by 2050, both below 2005 levels. A willingness to commit to double-digit reductions was positive, but none of the countries' targeted goals went beyond 2020.

Achieving these kinds of reductions by mid-century is quite simply, a Herculean task. It took decades of work for the EPA to finally lock in the first national greenhouse regulations on passenger vehicles, but even that breakthrough comes up well short of the trajectory required for an 80 percent reduction by 2050. Imagine that vehicle manufacturers continued to reduce greenhouse gas emissions by the same rate set in EPA's 2017–2025 program, 5 percent per year, all the way to the middle of the century. The 2050 model year cars would

average approximately 180 mpg. This sounds impressive, and is a necessary step toward meeting future carbon targets. But even introducing vehicles in 2050 that can drive from New York to Miami on seven
gallons of gas is not enough. According to the 2013 National Academy
of Science report, "Transitions to Alternative Vehicles and Fuels,"[2]
reaching the 80 percent emission reduction milestone would require
*all* cars on the road—not just the new ones—to get 180 mpg. This is
very hard to imagine happening.

Another measurement of how far just the global transportation
sector is from meeting these 2050 reduction goals appeared in a 2012
report by the International Council on Clean Transportation.[3] The
ICCT's survey of national policies adopted formally since 2000 found
that these regulations would only reduce global consumption of oil,
the main energy source for transportation, by 14 percent. Compared
to the pre-Obama administration status quo, current reductions are
encouraging, but without continuing improvements we will not meet
the targeted 80 percent reduction by mid-century.

What's more, the global population is expected to explode from
about seven billion today to over nine billion in 2050. These extra two
billion people, most of them in developing countries, will likely represent a huge increase in the number of drivers around the world. The
industry research group IHS Automotive forecasts that, by 2020, total
global vehicle production will exceed 100 million, versus just 80 million
today.[4] In large part because of all these new vehicles, many experts also
predict that petroleum demand will also skyrocket. Around the time our
first greenhouse gas rules were finalized in 2010, the world was using
about 87 million barrels per day. If that scenario played out unchanged
over the next few decades, the International Energy Agency estimates
demand to reach nearly 100 million barrels a day by 2035.[5] Dumping
that huge amount of additional carbon dioxide into the atmosphere
would be a climatic catastrophe—and make the mid-century goal of
reducing greenhouse gases by 80 percent all but impossible.

The only bright side to these dour numbers is that they use relatively static trends to predict the future. If nothing else, the next several

decades promise to be dynamic and unpredictable, with enormous changes in demographics, social trends, and technology. I believe that we can still radically transform the way humans transport themselves around the world and reduce greenhouse gases by 80 percent in 2050. But we first have to imagine what that world will look like twenty or thirty years down the road.

I have identified four future trends that I believe will determine what and how we will drive over the next few decades (Figure 2):

1)   The growth of urbanization and megacities
2)   The potential arrival of demand-side peak oil
3)   The international adoption and convergence of greenhouse gas regulations
4)   The social impacts of connected living and working

**Figure 2. The Trends That Will Shape the Car of the Future**

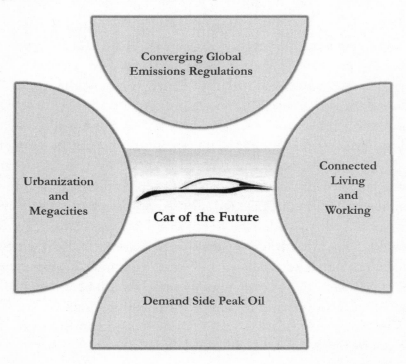

During President Obama's first term, my team at the EPA was able to use the confluence of a comparatively narrow range of tools, including a Supreme Court decision; years of engineering, policy, and legal work; the rise in gas prices; the leading role of California and other states; and the assistance of nongovernmental groups to seize a political opening. Today, individual states, nongovernmental organizations, industry, and regulators around the world must harness the many opportunities presented by these four future trends to create the cleaner and smarter transportation of tomorrow.

### Urbanization and Megacities

In 1350, following the ravages of the Black Death, there were about 370 million people on the earth—or a little more than quarter the population of China today. Ever since then, the world has experienced continuous population growth. During the past two centuries, the burgeoning demands of people on the planet have repeatedly led to predictions of eminent catastrophe. Perhaps the most famous of these dire pronouncements was Thomas Malthus's 1798 *An Essay on the Principle of Population,* which laid out what became known as the Iron Law of Population. Essentially, Malthus predicted that growing populations would always result in downward-spiraling wages—and an eventual collapse of civilization to a subsistence level survival. Later, as population reached unprecedented levels of growth following World War II, critics predicted that the earth would be unable to feed these multiplying numbers. Today many experts predict that the extra two billion or so people who will share earth in 2050 will accelerate our climate emergency, making greenhouse gas reductions nearly impossible.

While the numbers moving forward are certainly daunting, well-argued predictions on population catastrophe have proven wrong before. The Malthusian collapse, predicted to occur by the mid-1800s, was headed off by innovations in technology, economies, finance, and governance, among other advances. On average, people around the world have instead experienced a dramatic rise in their quality of life

since Malthus's prediction. Likewise, the Green Revolution, a collection of more productive agricultural management techniques first
introduced in the 1940s, has so far prevented the widespread starva-
tion predicted in the mid-twentieth century. Today, we'll need multiple advances, including new clean car technologies, low carbon fuels,
urban planning, and social change, to prevent another population-led
calamity. We'll also need these innovations to match the specific places
in which this growth will occur, dense urban areas.

★   ★   ★

In 1950, greater New York was the world's only megacity, a city with
10 million or more inhabitants (Figure 3). By 1970 the number had
doubled to two, with Tokyo joining the list. Today there are twenty-
three such cities, and in just ten years there will be thirty-seven massive, crowded, resource-sucking megacities, most of them in the
developing world.

The bad news is that, today, these urban monsters tend to produce
a disproportionate amount of the world's greenhouse gas emissions.
According to the UN-HABITAT's 2011 *Cities and Climate Change*
report, cities already account for up to 70 percent of harmful greenhouse gas emissions while containing slightly over 50 percent of its
population and occupying just 2 percent of its land.[6] The good news
is that they don't have to have such a negative impact. Urban planning,
including public transportation and other infrastructural investments,
can make a huge difference.

For example, I live in the Washington, DC, metro area. Greater
Washington is a thinly spread out mix of low-rise buildings, thick
traffic, and sprawling suburbs. My home is in Virginia, just a few miles
from the EPA in DC, but I commuted to work by car along with tens
of thousands of others in the area. In comparison, my daughter lives in
Manhattan, a densely packed city where people are much more likely
to live in smaller apartments and get around by walking, using public

transportation, or taxis. As a result, Washington, DC's annual per capita greenhouse gas emissions, approximately twenty tons of carbon dioxide equivalents, are nearly three times New York's roughly seven tons.[7]

This same phenomenon repeats itself again when we compare American cities as a whole to Europe. Cities in Europe tend to be more compact and have better public transport. As a result, people have lower car ownership and usage. Primarily because of planning, the contribution of European urban areas to climate change is even lower.

Because nearly all the population growth through the mid-century will be in cities, the way in which these areas develop will have a disproportionate impact in efforts to combat climate change. While we

**Figure 3. Past and Future Megacities of the World**

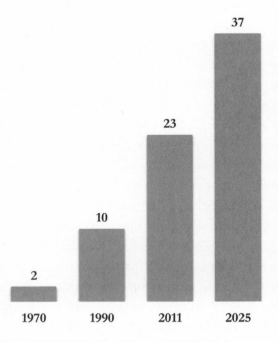

*(Source: UN World Urbanization Prospects, 2011 Revision)*

can't know exactly how successful cities will be at reducing green-house gases over the next few decades, we can be certain that urban areas will have enormous incentives to embrace radically cleaner and smarter urban planning. The health impact of air pollution and congestion, for example, is already a critical issue in the developing world's megacities.

★   ★   ★

At the June 2014 World Cities Summit in Singapore, Glynda Bathan-Baterina, the deputy executive director of Clean Air Asia, detailed some of the dangers floating in urban air globally. Speaking to many of the leaders of the world's megacities, BathanBaterina said that, according to a new study from the World Health Organization, seven million deaths worldwide resulted from indoor and outdoor air pollution-linked diseases, with about 2.6 million of those deaths in Asia. Aside from the human toll, this health crisis costs cities an enormous amount in lost worker productivity and healthcare.

That same year, the Organization for Economic Cooperation and Development (OECD) Secretary General's Office released a report detailing the huge role that transportation played in the rising health impacts of air pollution. Titled *The Cost of Air Pollution: Health Impacts of Road Transport*, the report estimates that outdoor air pollution is responsible for approximately 3.5 million deaths worldwide at a cost of $3.5 trillion (Figure 4). In the OECD countries, road transport emissions are estimated at about $850 billion, annually. China and India combined account for a comparable number of around $750 billion annually, for a world total of nearly $1.6 trillion.[8]

Without government action, the report makes clear that these costs will continue to rise. In other words, it is incredibly expensive for cities to ignore their transportation–related air pollution. Not surprisingly, some urban areas already have taken action.

To clean up its air, Germany's government has established  an "*Umweltzone*" in regions that exceed the pollutant emission threshold

**Figure 4. Deaths from Outdoor Air Pollution by Region (2005 and 2010)**

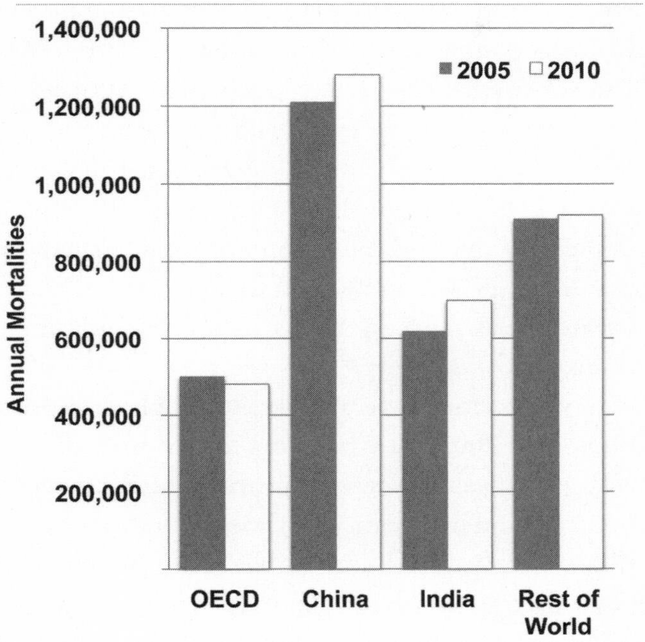

*(OECD, Institute for Health Metrics and Evaluation, 2013)*

set by the European Union. There are currently around fifty of these restricted "environmental zones" in Germany, and every car entering them must meet current European emission standards. Parking or driving in the zone without meeting the emission standard leads to a penalty of up to €40 ($54).

Paris set the ambitious goal of reducing carbon emissions from traffic by 60 percent in 2020. Among other measures, the city has reduced city parking spots by 9 percent and is charging for formerly free spots. As megacities struggle with air pollution, residents and local governments will likely also embrace emerging low and zero emission vehicle technologies.

★  ★  ★

With the enormous numbers of new cars predicted to hit their streets in the next decade, megacities are also focused on congestion. Traffic is not just dirty and frustrating; it costs large cities around the world hundreds of billions of dollars each year. According to Andreas Mai, Cisco's director of smart connected vehicles, in the United States "we waste more than 90 billion hours in traffic jams and generate 230 million metric tons of carbon."[9] The C40, an international network of sixty-three cities that collaborate on local climate actions, also highlights what a high proportion of this pollution is from inefficient transit. The group estimates that private motorized transport in cities accounts for less than a third of all journeys but is responsible for more than 70 percent of emissions.[10]

As a result, some cities have also instituted plans to reduce automobile traffic. Since 2003, for example, London has had a congestion charge of £10 ($16) for cars driving in the central city on weekdays between 7:00 a.m. and 6:00 p.m. The money is not just a disincentive to drive; it's also invested back into public transportation. In addition to penalties for driving in urban centers, congestion will also accelerate the development of "smarter traffic." Vehicles linked with other vehicles and traffic control systems would be able navigate cities much more efficiently than drivers currently do.

The growing importance of megacities in the future of transportation has not been lost on forward-thinking car companies. When I spoke with Ulrich Kranz, a senior vice president of the electric BMW i Product Line, he explained that development of the company's electric car began in 2007 when his team "started visiting different megacities of the world in Asia and the U.S." Among the most important feedback was that "people really want to keep their individual mobility." And "they are also really conscious of the environment," both of which were factored into designing the BMW i series.[11]

★   ★   ★

Beijing has become a poster child for both unhealthy environmental pollution and gridlock in developing world megacities. Over the past twenty years, my team and I visited the city numerous times to provide technical assistance to China's Ministry of Environmental Protection on transportation emission control policies and regulations. We have also hosted Chinese scientists and engineers at our Ann Arbor lab for training on vehicle emissions testing and regulatory analysis. In Beijing, the pollution was so bad that I could rarely see the sky. There were days when nearby office buildings disappeared behind the smog. On the occasional clear day, our Chinese colleagues joked that they ordered the clean air just for us.

But there was a silver lining to this ever-worsening smog. In January 2013, the air quality plummeted so low—even by Beijing's standards—that the country was forced to take action. On January 12, the city's pollution registered at twenty times worse than the World Health Organization's minimum safe levels. Finally, after more than a decade of negotiation, support from Mike Walsh and his brainchild, the International Council on Clean Transportation, among other groups, and technical support from my EPA team, China instituted strong fuel quality standards to spur super-low emission vehicles. The measures include a target to reduce sulfur to 10 parts per million by 2017. Particulate emissions, or soot, will also be reduced between 10 and 25 percent, depending on the city. Hundreds of cities across China now have expanded air quality monitoring to track progress. With continued strong measures like these, Beijing could reverse its fortunes relatively quickly.

The dismal situation in Beijing is, in fact, reminiscent of mid-century Los Angeles. In California, aggressively confronting its problem of horrible air pollution quickly made the state a world leader in clean vehicle technology. The Chinese have taken notice of the similarities between their largest city and 1960s Los Angeles and studied how California successfully dealt with its smog.

Beijing's horrible air pollution has actually played a role in jump-starting the Chinese New Energy Vehicles Program, which promotes the development and introduction of vehicles powered by electricity. Reversing the current trend, an overcrowded megacity could actually play an important role in reducing pollution and greenhouse gas emissions worldwide. Just as in California, efforts to clean up Beijing—and perhaps all of China—could result in the world's largest country becoming a center for electric and hybrid vehicles. But cleaning up Beijing's urban air pollution isn't just a matter of embracing better technology.

★   ★   ★

In 2012, the Paulson Institute, a group that promotes economic growth and environmental preservation in the United States and China, offered an action plan for Beijing that included many suggestions for improvements in urban planning. Among transportation-focused advice, authors Hal Harvey and Chris Busch recommended a departure from the sprawling mega-block development common in many Chinese cities in favor of smaller blocks with more mixed-use streets. Development that places various businesses and services next to residential zones creates walkable neighborhoods, where residents can work, shop, and socialize without hopping into a car every time they need to do or buy something. For longer trips, China is already building thousands of miles of subway lines, but the authors argue that cheaper rapid transit bus lines would be more effective.

The initial measures by China to clean up the air and mitigate congestion in Beijing are hopeful signs that mayors and governments of all large urban centers and megacities will be compelled by their own pollution to take serious, transformative action. The solutions proposed for Beijing will likely apply to many of them, but specific local innovations to reduce traffic and pollution will also be extremely important. In some cases, urban regulations will include explicit limitations on greenhouse gases. Any efforts to combat these problems will have a positive impact on awareness and future reductions of greenhouse gas emissions as well.

Of course, the best-case scenario for the future of transportation in megacities goes beyond cleaning up or better managing traffic. Singapore discourages its five million residents not just from driving but from car ownership itself. The city holds auctions for a limited number of vehicle registrations. These Certificates of Entitlement exceed the value of most cars, with the net effect of more than doubling vehicle purchase costs. The result is widespread usage of Singapore's high quality public transit and taxi services. When rapidly growing, polluted, and traffic-snarled megacities look at the overall "transportation system," they may come to the same conclusion as Singapore: private vehicles simply do not fit into the future. If burgeoning megacities like Sao Paulo, Mumbai, and Nairobi take similar action in the future, large urban centers will be taking a huge step toward greenhouse gas reduction targets.

### Demand-side Peak Oil

In 2011, Don Clay, a former EPA executive and one of my EPA mentors, set up a meeting for me at the Koch Industries' headquarters in Wichita, Kansas. Koch Industries does business in the petroleum, chemical, fertilizer, paper, and ranching industries, all sectors that frequently come in to conflict with the EPA's efforts to clean up the country. The business' owners, David and Charles Koch, are also synonymous with conservative and libertarian efforts to disrupt, block, and destroy environmental measures to reign in global warming.

I didn't understand Don's decision to work for the Koch brothers, but we had stayed in touch over the years. By 2011, I was working on what were called the Tier 3 auto standards, a program designed to clean up the sulfur in gasoline and reduce car emissions. I was meeting with the oil refiners to better understand their capabilities to reduce sulfur in gasoline. Some of my EPA colleagues were pessimistic, but I thought that talking to all stakeholders, however antagonistic, was the only way to create good regulations.

At the headquarters, I met Koch Industries' executive vice president of operations, Jim Mahoney, and did my presentation on our proposed sulfur reduction regulation. Afterwards, he said that EPA's upgrades would cost too much and that he couldn't support them. His response was hardly a surprise, but I told Jim we should stay in touch. "Maybe we could still reach an agreement."

He shook his head. "Margo," he said, "it's not the Tier 3 issue. It's really about your greenhouse gas standards for cars in 2025."

I didn't like Jim's answer. The greenhouse gas standards were one of my proudest accomplishments at EPA, not to mention critical for the future of life on earth. I was also disappointed that his company wasn't even going to consider the Tier 3 rules because of a totally separate agreement we had negotiated with the auto industry. But I will give him credit for being straightforward. He was probably the only person in the oil industry who was forthright with me on this issue.

There is an enormous difference in how oil companies view environmental regulations to clean up traditional pollutants and the new ones aimed at reducing green house gas emissions. Most of the time these corporations did not support EPA regulations to make their products cleaner, but after the rules became binding they invested in new technology and ended up selling a better version of their product at a profit.

Greenhouse gas standards are different. Globally, transportation is almost entirely dependent on petroleum-based fuel. As a result, today there is only one way to significantly reduce the greenhouse gas emissions from cars and trucks and airplanes: burn less oil. Unlike previous environmental measures, greenhouse gas standards won't compel oil companies to improve their product—they will displace it. Our standards through 2025 will create markets for all sorts of technology, from more efficient gasoline-powered cars to redesigned electronic architectures to cheaper batteries, but all of them use less of the petroleum companies' product. Jim's terse but honest reply said it all. The oil industries view any regulation that limits greenhouse gas emissions as an existential threat.

★   ★   ★

What's curious about this concern is that many groups, including oil companies themselves, predict global oil demand to go through the roof over the next few decades. For example, the US Energy Information Administration's 2013 Energy Outlook predicts total energy consumption for transportation in OECD nations will decrease by an average of 0.1 percent per year through 2040.[12] However, non-OECD demand for transportation energy use will nearly double in that same time period. China and India will show average annual increases of 4.1 and 4.6 percent, respectively, in energy use per capita. The net result is that energy consumption in the global transportation sector rises from a 2010 baseline of about 101 quadrillion BTUs to about 140 quadrillion BTUs in 2040, as shown in Figure 5.

This devastating trend toward increased energy use in the transportation sector is almost entirely due to a rapid growth in vehicle sales, primarily in the developing world. For example, Pricewaterhouse Cooper's Autofacts forecasting group suggests that the developing Asia Pacific region will increase production 61 percent from 2013 to 2020, almost entirely to satisfy regional demand.[13] Additionally, the region will import another several million vehicles from Europe and North America.

China alone has already eclipsed the United States as the world's largest market for light duty vehicles and has enough roads and related infrastructure to accommodate twice as many vehicles. In 2014, Chinese sales were expected to grow to 24 million, three million more than just the previous year. By comparison, US total light duty vehicle sales were a little less than 16 million. If this growth were to continue, the Chinese market could surpass 30 million sales a year, a larger number than North America plus Europe in their heyday years, combined. It is based on stratospheric numbers like these that oil consumption is predicted to grow so rapidly.

There are, however, reasons to doubt the projections for rapidly rising oil demand that are endorsed by energy agencies and oil companies. In fact, rather than exponentially increasing, I believe that

**Figure 5. Transportation Sector Energy Consumption (Business-as-Usual Scenario)**

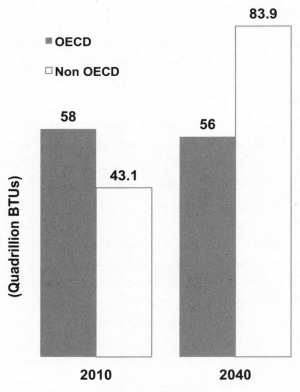

(Source: EIA International Energy Outlook 2013)

demand for oil will begin to fall globally before 2050. A wide range of experts are also predicting or advocating for similar trends.

★   ★   ★

On March 26, 2013, Citibank's commodity research division released a document titled "Global Oil Demand Growth—The End Is Nigh." A few months later, the *Economist* took up a similar argument in the editorial "Yesterday's Fuel."[14] For both groups—neither of them run by utopian environmentalists—the scenario leading to oil's decline was similar.

A combination of recently discovered, relatively cheap natural gas and rising fuel economy mandates was pushing the world toward "peak oil." But instead of the traditional theory of peak oil in which the world's supply can't keep up with demand, this would be a demand-side peak.

"In the rich world, oil demand has already peaked: it has fallen since 2005," wrote the editors at the *Economist*. "Even allowing for all those new drivers in Beijing and Delhi, two revolutions in technology will dampen the world's thirst for the black stuff." The two technologies the authors refer to are fracking, whereby abundant and cheaper natural gas is likely to replace oil in certain applications (e.g. ships), and much more efficient internal combustion as well as electric and hybrid powertrains will be available. Because oil is far and away the leading source of greenhouse gas emissions in the transportation sector, it is important to understand how "black gold" has begun to lose its luster with a wide array of people, from commodity experts to economists to national security experts.

★　★　★

In the summer of 2013, former CIA director James Woolsey told me a story about the beginning of his personal commitment to freeing the country from oil dependence.[15] In mid-October 1973, when he was the general counsel for the Armed Services Committee, Woolsey had to stop for gas on his way to work. He was in a hurry. That day he was supposed to run a Congressional hearing on the recent outbreak of hostilities between Israel and Syria and Egypt known as the Yom Kippur War. But because of pressure from the Arab oil-producing nations, gas prices had just begun shooting upward amid fears of shortages. Woolsey missed most of the hearing waiting in the long line at the pumps. "I turned pretty hostile to oil then," he said, "and that was forty years ago."

Woolsey had grown up in the heart of oil country, Tulsa, Oklahoma. His father "was a lawyer who sometimes represented oil companies and sometimes represented people who sued oil companies."

But Woolsey became an early advocate for alternative fuels including cellulosic ethanol. Because he is not seen as a traditional environmentalist, Woolsey has also had some opportunities to get the most ardent climate change deniers to understand the other benefits to measures that would reduce global warming.

A few years ago, Dana Rohrabacher, a conservative congressman from Southern California, was digging into Woolsey during a committee hearing on climate change and national security.

"Dana was really hard," said Woolsey. "He just had a few seconds left, so I said, 'Congressman, let's hold up here just a second and set aside climate change. Do you realize that all four things that I just proposed would make us less vulnerable to terrorist attacks and oil cutoffs?' He said, 'Oh, well, if you doing it for that reason, it's fine.'"

★   ★   ★

Because it is just about the only fuel that transports people and goods around the country, oil has been a prominent national security concern for much of the twentieth century. Unfortunately, much of the oil imported in the United States comes from the Middle East, making the uncertain politics of countries 6,000 miles away a prominent concern in Washington, DC. Because of this dependency, the United States continues to be dragged into countless regional conflicts.

Although securing Middle Eastern oil through whatever means necessary had already been the unofficial policy of the United States for decades, President Jimmy Carter's January 1980 speech laid it out in plain English. "Let our position be absolutely clear: An attempt by any outside force to gain control of the Persian Gulf region will be regarded as an assault on the vital interests of the United States of America, and such an assault will be repelled by any means necessary, including military force." What became known as the Carter Doctrine has since been the officially endorsed mainstay of the US national security policy.

President Carter's speech was forceful, but oil's virtual monopoly as a transportation fuel has left the nation hobbled. Alternately protecting and invading the largest producers of oil in the Middle East has resulted in great sacrifice by Americans. The United States has led invasions of the Persian Gulf each of the past two decades and is still trying to extricate itself from an increasingly messy situation in Iraq. Thousands of Americans—both military and contractors—have lost their lives in these actions. Additional tens of thousands of troops have been wounded in action or evacuated due to injury or disease. Hundreds of thousands have filed disability claims for both physical and psychological damage.

Protecting these sources of oil in terms of direct cost war spending alone—i.e., tax dollars—is immense. Brown University's Cost of War project estimates that military operations in Iraq, Afghanistan, and Pakistan have cost the United States nearly $3.4 trillion from fiscal year 2001 through 2014.[16] Adding the cumulative interest on past Pentagon and State/USAID war appropriations, the number could easily double. Were the United States not so dependent on oil, lives could have been saved and much of this money could have been invested in schools and infrastructure—or alternative fuel development. But the United States' dependency on oil also draws criticism for a different way in which it damages the nation's and the world's economy.

★  ★  ★

For decades, people have explained away the toxic clouds from burning fossil fuels and foreign wars associated with oil by claiming that they were the unpleasant price of progress and growth. Carmine Difiglio, the deputy director for energy security at the Department of Energy—and someone with whom I worked closely—draws the opposite conclusion.[17] According to Difiglio, our global dependency on oil may actually be limiting economic growth. In his analysis, oil's unique position in the worldwide transportation sector has a negative effect on the global economy because of what is known as price elasticity.

Price elasticity refers to the cost increase consumers are willing to pay before switching products. For example, if the price of corn syrup is too high, Coca-Cola bottlers could switch to sugar. If the cost of Coca-Cola gets too high, people could buy Pepsi. Even huge utilities have some degree of flexibility in electricity generation. If the price of coal is too high, they might buy power from a different, cheaper source like nuclear, natural gas, solar, or wind. This dynamic acts a sort of market-based price control. But when gasoline and diesel prices go up, there is nothing to switch to, a unique situation for such a huge commodity. Because of this monopoly position, economists say oil's price is very inelastic.

Faced with higher oil prices, people and companies don't have the option of using something else to power their vehicles. Commuting workers, truck drivers, airlines and others have always had to simply pay the higher price for gas or diesel or jet fuel. As a result, rising fuel costs will eventually begin to eat into the purchase of other goods and services, creating a drag on the economy. According to Difiglio, the most extreme rises in the cost of oil, called price shocks, create such huge negative impact on the economy that they quickly lead to huge drops in global GDP.

Looking at data from 1970 to 2010, Difiglio illuminates just how dangerous oil's monopoly position is. Over that forty-year period, every oil price spike was followed by GDP tanking worldwide. Even more significantly, there were no major sustained drops in global GDP that weren't preceded by spikes in oil prices.

The 1973 oil shock, which had made an early alternative fuels convert out of James Woolsey, was an enormous blow to the world's economy. That year, global GDP growth was an impressive 6 percent. The Arab oil-producing states' embargo began to push oil prices up at the end of the year before they spiked in 1974. That year, world-wide GDP plummeted to just above 2 percent, a third of what it had been the previous year. In 1975, the growth dropped further to 1 percent. Ever since then, the world has faced repeated oil price shocks, usually created by events in the Middle East, with similar economic downturns.

In 1979, just four years later, global GDP had climbed back up to 4 percent. Then the shah of Iran was deposed and oil prices suddenly doubled. Over the next two years, GDP growth remained at 2 percent, before falling again to 1 percent in 1983. Seven years later, the 1990 Gulf War spiked oil prices, dropping GDP growth from around 3 percent to at or below 1.5 percent for the next three years. Following a brief economic downturn in 2001, worldwide economic growth sat at around 4 percent for the next five years and seemed immune to both the epic rise in oil prices and the ongoing war in Iraq. Then between 2007 and 2008, there was sharp $40 per barrel rise in the price in oil. In this case, it was the banking crisis in 2009 that triggered the worst recession since the 1930s, but demand for oil had already begun to slacken before that as the world's economy had started slowing. In 2009, the global economy actually contracted by 2 percent.

The only ways to resolve the economic and energy security problems caused by the oil monopoly are car technologies and fuels that introduce competition to the transportation fuel market. But this is not just a domestic problem. Countries around the world have also been devastated by oil price shocks—some much worse that the United States—and held hostage by supply issues. Although these nations may not be losing lives in oil wars, countries from Spain to China have every bit as much interest in energy security. As a result, a broad and international coalition of economic, environmental, and energy security imperatives will drive the further development of low carbon fuels and electric cars. More important, vehicles that provide some of those advantages are already available and dropping in price. This synergy between a wide range of interests in reducing the dominance of oil in the transportation sector along with the commercial emergence of cars, trucks, and fuels capable of doing that will exert downward pressure on the demand for oil.

In fact, these trends seem to have already impacted the estimates of the International Energy Agency, the Paris-based coalition of twenty-nine member countries that provides leading policy advice on energy and oil-related issues. Citing a combination of high oil prices, environmental concerns, technology advances, and other factors, the agency adjusted its 2014 *Medium-Term Oil Market Report* to reflect a slow-down in global oil demand. Beginning in 2019, an "inflection point" is likely to include "fuel-switching away from oil and conservation measures [that] will likely blunt the demand impact of economic and population growth, causing oil consumption growth to decelerate."

The basics of supply and demand in the oil market will have a huge impact on the development of future car technology. If oil demand does continue to rise unabated, then it is going to be harder to remove the added greenhouse gases from a global vehicle fleet that the US Department of Energy predicts will grow from one billion today to 2.5 billion by 2050,[18] especially if the majority of those new vehicles rely on largely conventional internal combustion technology. Of course, a steady rise in demand would also keep oil prices high, incentivizing consumers to adopt fuel efficient, low carbon vehicle technologies.

If, as seems increasingly plausible, oil experiences a peak in demand, carbon dioxide emissions should drop in tandem with reduced oil consumption. Unfortunately, a reduced demand will likely also cause oil prices to drop. That in turn will make petroleum-based fuels cheaper, making alternative fuels and drivetrains less competitive and probably disincentivizing their development and adoption. But either way, oil is unlikely to occupy its century-old monopolistic position in fueling transportation, opening up space for alternatives.

### International Confluence of Greenhouse Gas Regulations

In June of 2001, I walked into the reception area of the Bellagio Center, a Rockefeller Foundation retreat set on the shore of Lake Como in the rolling hills of Northern Italy. A collection of buildings

dating as far back as the fifteenth century, the center's campus has provided a peaceful environment for the residencies of artists, musicians,
policy makers, and numerous Nobel laureates and Pulitzer Prize winners. Across the lake from the center's grounds, Cyprus trees and green
slopes give way to the snow-covered beginnings of the Alps.

The receptionist put me in a villa in which she said President Kennedy once stayed. Back home, the environmental politics of the Bush
administration had already turned ugly. The current politicized situation, I thought, is far from Kennedy's 1961 inaugural rallying cry: "Ask
not what your country can do for you, but what you can do for your
country." This was about the last few minutes I had for introspection
at the Bellagio Center.

I was not on vacation but meeting with seventeen other transportation air quality policy makers, academics, and experts from Europe,
China, India, Japan, Mexico and the United States, including one of
my predecessors at the EPA, Mike Walsh. Hal Harvey, the then CEO
of the Energy Foundation, and Charlotte Pera, also of the Energy
Foundation, had brought us all together for three intensive days of
discussions. The meeting was prompted in part by the realization that
while the auto and fuels industries were global, the policies governing
them were not. Our goal was to agree upon a common set of goals
and practices for vehicles and fuels.

Three exhausting days later, we produced what was called the Bellagio Memorandum, "a consensus of all 18 participants on over 40
principles—or guidelines—that, taken together, should constitute
the policy future for motor vehicles and transportation fuels. These
principles can form a clear, explicit guide for policymakers around
the world and for automakers and oil companies as they design their
products for the next decade."

Shortly thereafter, this council was formalized into an organization called International Council on Clean Transportation. Today,
ICCT provides technical and policy support to many governments
including China, India, South America, Europe, and the United States.
The group's enthusiastic executive director, Drew Kodjack, and his

team have managed to apply the principles first established in Bellagio across the world for the past fifteen years, providing technical and policy support on transportation emissions standards, fuels, and technologies. The result of this international collaboration has been effective in the battle to reduce greenhouse gases globally.

The 2009 Copenhagen Summit was an example of the limited success that international attempts to set enforceable greenhouse gas reductions have faced. But the largest nations in the world have managed to converge on one specific goal: adopting tough standards on vehicular emissions. As in the United States, these regulations generally started with mostly visible pollution like airborne particulates or smog before graduating to greenhouse gases. Most recently, China joined the other three biggest economies in the world, the European Union, United States, and Japan, in setting aggressive fuel economy targets. Not only are these countries the world's biggest economies by GDP; they are also far and away the largest vehicle manufacturing nations and represent the largest market for auto sales. Many other nations in the developing world are also following by phasing in greenhouse gas emissions and fuel economy standards and translating their targets into actionable regulations.

According to the International Council on Clean Transportation, some form of fuel economy and greenhouse gas emissions standards are now in effect for more than 70 percent of new vehicles and their markets globally. Although each of the historical starting points for the countries were fairly disparate, their standards and targets will all converge between 2020 and 2025 (Figure 6). Aside from their obvious benefits for air quality in individual nations, the convergence of these standards provides an additional incentive for automakers to develop advanced greenhouse gas reduction technology.

★   ★   ★

Aside from simply forcing car companies to produce cleaner cars, the convergence of these regulations has an added benefit. During

**Figure 6. GHG Reduction Standards and Targets for Passenger Cars in Key Countries (2000–2025)**

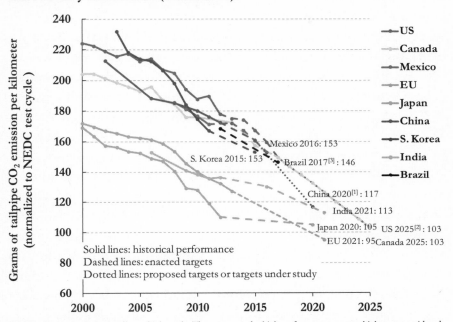

[1] China's target reflects gasoline vehicles only. The target may be higher after new energy vehicles are considered.

[2] US fuel economy stadards set by NHTSA reflecting tailpipe GHG emission (i.e. exclude low-GWP refrigerant credits).

[3] Gasoline in Brazil contains 22% of ethanol (E22), all data in the chart have been converted to gasoline (E00) equivalent

[4] Supporting data can be found at: http://www.theicct.org/info-tools/global-passenger-vehicle-standards.

*(Source: ICCT)*

negotiations with automakers in 2009 and 2011, one of our most important points of leverage was the car companies' desire for a unified national market. Ever since Henry Ford introduced the production line, the big automakers' business model has been entirely contingent on maximizing scale, cheaply producing millions of the same model for a single market. This model allows them to turn a profit while selling their vehicles at a price attractive to mainstream consumers.

In 2009, the automakers knew that if California and a dozen or so other states adopted different emissions requirements, their mass-produced vehicles would lose close to half their national market.

Producing two different cars for two different markets was possible, but it could reduce car companies' efficiencies of scale and, as a result, their profits. We offered the companies a more profitable, unified market in exchange for higher fuel efficiency and reduced greenhouse gas emissions. The automakers were willing to sell cleaner, slightly more expensive cars if the entire United States could buy the same models. The convergence of international greenhouse gas standards between 2020 and 2025 will provide a similar unified market, but on a global scale.

As the Chinese, European, Japanese, and American markets require increasingly similar vehicle emissions standards, automakers can sell their cleaner cars to an even larger market. Companies can achieve the massive scale efficiencies of producing "world cars" that can meet multiple standards globally. This, in turn, has a dramatic impact on the market competition to reduce greenhouse gas emissions. As the auto manufacturers have realized that developing affordable and popular clean vehicles will give them a global edge, they have been more willing to invest and innovate in this technology. This expanded investment is, in turn, breeding new players and competition.

★   ★   ★

Chinese automakers have at least one specific hope for these regulations. To date, US emission and safety requirements have been a barrier to entry into the American market for China's huge car companies. But, as global fuel economy and tailpipe emissions regulations converge, these auto manufacturers are trying to leverage these changing dynamics to gain a foothold in the American market. By building a "world car" for both the American and their domestic markets, China's car companies will benefit from the huge economies of scale offered by the world's two largest auto markets. Geely, the parent company of Volvo, announced in June 2014 that some of Volvo's cars sold this year in the United States will be manufactured in China.

The Chinese strategy of using increased regulations on vehicles to gain market share has a successful historical precedent. As fuel consumption and environmental requirements went into effect in the 1970s and '80s, the American auto industry made fighting the fuel economy and tailpipe emissions regulations their primary mission. The American automakers' misguided quest not only failed, but it helped guarantee the success of the Japanese auto manufacturers. Car companies like Toyota, Honda, and Nissan produced more popular fuel-efficient cars equipped with catalytic converters than their American competitors. Within a decade, all of the companies were major players in the US market. In 2007 Toyota even eclipsed GM as the world's largest automaker.

Whether or not China can repeat this success is unknown, but it is certain that this global confluence of greenhouse gas regulations will accelerate the global race to produce the world's cleanest cars, including advanced hybrids and pure electric cars.

### Connected Living and Working

In 2012, 2.4 billion people—one in three people on earth—had Internet access. Put another way, the number of people digitally connected today is greater than the populations of North America, South America, and Africa combined. This represents a 550 percent increase in Internet access since 2000 and a sea change in global connectivity. The connected world was, until recently, the province of citizens living in wealthy OECD nations and the elite in the developing world. Today, social networks span the world and transcend lines of class, race, and ethnicity. Facebook alone has over 1.35 billion users globally. While the world is rapidly becoming more crowded, digital connectivity is accelerating at a much faster rate. These new communication dynamics will have dramatic effects on the future that match even those of population growth.

The first generations to grow up in this era of immersive connectivity have already generated enormous social and political transformations. In a May 13, 2014, op-ed, *New York Times* column-ist Tom Friedman coined the phrase "the Square People" to refer to the young generation of Tahrir Square in Cairo, Egypt, and the Maidan Square in Kiev, Ukraine, who have been at the center of world-changing political activism. The Square People represent a new kind of political movement that communicates, self-organizes, and acts in unison using Twitter and Facebook to bypass traditional media channels. But the impact of connected living goes far beyond head-line-grabbing political events, reshaping a multitude of economic and social trends, including transportation.

★   ★   ★

As with any fast-developing social and political trend, predicting the future impact of connected living is an uncertain science. But examin-ing its effect among young people in the nations where Internet access has been established longest and is the most extensive is a good place to start. In these wealthy countries where most twenty-five-year-olds can't remember a time before the Internet, a different hierarchy of val-ues has emerged. For example, a recent study showed that thirteen-to seventeen-year-olds in both the United States and the United King-dom would much prefer to give up an allowance from their parents or going out to eat, to the movies, or to sporting events than relinquish their Internet connections. For teens, access to a technology that was barely used a few decades earlier has supplanted these other traditional pleasures. These high schoolers most want to stay plugged in and com-municate with their friends and the world electronically. This, in turn, has a huge impact on how they view transportation. (See Figure 7.)

The revolutionary social impact of cars in the twentieth century was their automobility, the ability to be where you wanted, when you wanted. Instead of relying on walking or scheduled trains for transport, people could suddenly travel spontaneously. The American romance of this freedom was popularized in songs like the Beach Boys' "I Get Around" and literature like Jack Kerouac's *On the Road*. From *Easy*

**Figure 7. Reluctance to Disconnect (US & UK Age 13–17)**

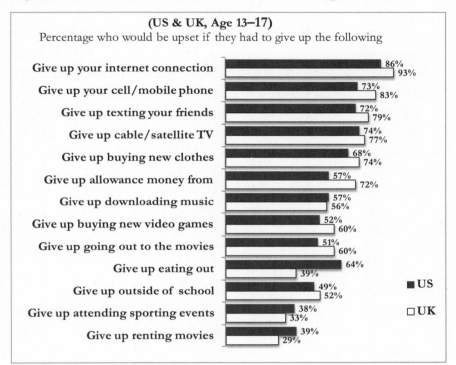

(Source: JWT Gen Z: Digital in their DNA, APRIL 2012)

*Rider* to *Planes, Trains and Automobiles*, the road movie also became a popular subgenre. At drive-in theaters, you could even watch those movies in your car. But the growth of Internet-enabled connected living has both undercut and reimagined the previous century's version of "automobility." With a smartphone and Netflix, there's no need to drive to a theater.

My oldest daughter, Nicole, was in a transitional generation. When she turned sixteen, the only thing she wanted was a car. Though she could walk to school in about three minutes, it was not a cool thing to do. She refused to walk or bike, and she took the car. Despite my pleas, she was absolutely insistent on driving a tenth of a mile to school. Then one day she had a conversion. I was at work, not feeling very well, when my assistant Gladys Bryant interrupted me to tell me that Nicole was on the phone and wanted me to know that she had done her part for the environment and walked to school. She knew that this would make me feel better, and it did. Today, many of the sixteen-year-olds I meet don't want to drive, some for environmental reasons, but most because they have other ways to be connected.

Statistics reflecting young peoples' slumping interest in vehicles highlight this new definition of mobility. In Europe and the United States, more teens and twenty-year-olds are delaying getting their driver's licenses and often not bothering to get one at all. As recently as two decades ago, 87 percent of nineteen-year-olds had a driver's license; today that number is less than 70 percent. This phenomenon is already having an impact on American driving habits.

★   ★   ★

Last year, when I had lunch with former NHTSA administrator David Strickland, he described another related trend among young people: not only were they less interested in driving cars, they didn't want to own them either.

"People in this universe don't want to buy things," said David. "They don't own things. They own data. They've never bought a cassette, they've never bought a record, and they've never even bought a CD or DVD. Everything is streamed, everything is portable, and they like the notion of 'I only get to use something when I need it, and I don't have to have the responsibility of dealing with it when I don't.'"[19]

Led by the digitally-connected teens and people in their twenties, America's twentieth-century love affair with cars seems to be waning. As David pointed out, the country is already experiencing a significant change in car ownership. The average number of cars owned by a household is down to less than two cars, about the same number as two decades earlier. The overall per capita vehicle miles traveled in the

United States dropped again in 2013, making it the ninth consecutive year of decline.

Low-paying jobs, expensive gas, and environmental concerns have certainly helped push down these numbers. As the economy improves and young people pay off their student loans and find better jobs, it is possible that Americans will start driving more. For now, though, companies offering car ownership alternatives are betting that these trends are real and lasting. In 2000, Zipcar began a membership-based service that allowed people to pick up cars parked around town to drive on an hourly basis. Soon the market became crowded with competitors, including Car2Go and Hertz OnDemand. In 2013, Avis bought Zipcar in order to be more competitive with Hertz's on-demand service.

Another car ownership alternative, ride sharing, has taken off in Europe. Munich's Carpooling.com, for example, allows users to book available seats in private cars. It has been so successful that Daimler-Benz invested $10 million in mid-2012. The French ride-sharing company, BlaBlaCar, claims that it moves about 600,000 passengers per month across Europe. By comparison, the Eurostar train that connects London with Brussels and Paris moves about 900,000 passengers a month. Uber, a San Francisco–based company that makes apps to connect drivers and passengers, was recently valued at $18.2 billion.

Connected living is primed to impact the future of transportation in two ways. The first is technological: the development of affordable semi-autonomous and driverless cars. These vehicles could provide the ultimate on-demand service: transporting unlicensed or car-less passengers for the same price as picking up a Car2Go—and well below that of a taxicab.

The other change is social but could be even more dramatic. If the global spread of connected living encourages more and more people to use on-demand and shared vehicle services instead of owning cars themselves, the number of cars on the road will decrease rapidly. A study by the consulting firm Alix Partners found that every new

shared vehicle added to a fleet displaces as many as thirty-two vehicle sales.[20] The expansion of these trends worldwide is a nightmare
scenario for the auto and oil companies but would be a tremendous
victory in efforts to reduce greenhouse gas emissions.

Recognizing the importance of this trend, car companies are preparing for a future in which people are reluctant to become car owners. During our talk, Ulrich Kranz, the senior VP of BMW's i Product
Line, talked about a category of "mobility solutions," saying, "Japanese kids around eighteen to twenty are not getting driver's licenses,
because really good public transportation is available. We think we can
steer people into car ownership with an i3, but also we can introduce
the electric i3 into a car-sharing program. It will be appreciated by
younger people."

★   ★   ★

It is these future trends—the response of megacities to environmental pollution and congestion; a diminishing demand for oil beginning in the OECD nations before spreading to developing countries;
an historic convergence of international greenhouse gas regulations
on vehicles; and, last, connected living and a revaluation of personal
mobility—that will shape personal transportation options through
2050. All future efforts to reduce greenhouse gas emissions will take
place in a space defined by these four drivers. While I don't pretend
it will be easy to get to the 2050 goal of an 80 percent reduction in
greenhouse gases, I am sure that it is achievable in the context established by these future trends. I am also certain that the EPA must play
a role in getting us there.

For one, the EPA has been in a similar situation before. When
Richard Nixon created the agency, vehicles had already been spewing
deadly pollution for decades. America's cities were inundated with
smog and toxins. Then, between 1970 and 2010, new cars, pickup
trucks, SUVs, heavy-duty trucks, and buses became over 95 percent
cleaner for common but hazardous pollutants like hydrocarbons,

carbon monoxide, nitrogen oxides, and particle emissions. Sulfur in gasoline and diesel have been reduced by over 90 percent. The country is remarkably clean compared to the early 1970s. While our skies are not free of all pollutants, smog no longer obscures buildings in Manhattan or blocks out the sun in Los Angeles.

The EPA has also already played a leadership role worldwide. Leaded gas was first eliminated in the United States; now it's virtually nonexistent worldwide. Catalytic converters were first mandated in the United States; now it's nearly impossible to find cars without them. The challenges of global warming are daunting, but the huge breakthroughs of the previous forty years are exactly what happens during a clean transportation revolution.

Future cars will have to be even cleaner, smarter, and safer. They will have to contribute to the lower carbon footprints of urban centers and megacities. They will need to plug in to an increasingly connected world. Future cars will be "mobility solutions" that offer personal transport freedom. They will have to be at least as safe what we drive today. On top of all that, they will also have to excite consumers.

Tomorrow's cars will be the result of massive technological innovation, but we can't understand the future car simply by what propels the vehicle but what it runs on, what it is made of, and who drives it. All these basic elements of the automobility "system" will change irrevocably over the next few decades.

The transformation has begun with powertrains that offer low and even zero-emissions in order to meet the globally converging regulations. A wide range of such vehicles, from hybrids to plug-in electrics and fuel cell vehicles is already in dealership showrooms.

At the vehicle level, advances in strong, lightweight material technologies will help make any powertrain, existing or new, deliver better efficiency, as lower mass requires less energy to move. Alternate materials like high-strength steels, aluminum, carbon fiber, and magnesium are already playing a bigger role in making cars lighter. Where fuels continue to be used, for example in hybrids, they will need to be low carbon fuels to reduce carbon dioxide emissions.

Finally, autonomous cars that drive themselves will allow socially connected drivers to multitask even more than they do today—and much more safely. Combined with vehicle sharing, another trend among younger generations, future drivers will use cars differently than generations in the past and likely fewer of them. All of these dynamics will converge to create the future car.

Despite its emphasis on technology, this revolution will be different in a very important way. To succeed, we need to bring the world along with us. In the past, the EPA has been a world leader, but we will need to coordinate our actions with other countries even more closely.

This is no easy task. It will be a long, hard trek. But, fortunately, some of the new technologies that will help us fight global warming are already on the road. The following chapters examine some of the key technological changes that we must harness to meet our goal of 80 percent reduction in greenhouse gas emissions by 2050. As we push forward, we will also see the outline of the future car, revolutionary personal transportation that reduces greenhouse gas emissions and combats global warming.

# 15

# GOING FOR ZERO EMISSIONS

Since the 1970s, the powertrain has been at the center of the automotive industry's efforts to reduce emissions and improve fuel efficiency of vehicles. The automotive industry defines the powertrain as the systems that include the engine, motor, transmission, driveline and axles, as well as the electronic controls that help the system operate in an integrated fashion.

Powertrain engineering successes have produced ever-cleaner emissions. These improvements have traditionally targeted reductions in conventional pollution like nitrogen oxides, hydrocarbons, carbon monoxide, and particulate matter before very recently focusing on greenhouse gases. From traditional internal combustion engines to the new generation of electric or fuel cell vehicles, each powertrain alternative operates with its own specific requirements and limitations. An internal combustion vehicle, for example, will never achieve zero emissions. On the other hand, a fuel cell vehicle can get zero emissions but needs to be fueled by hydrogen from renewable sources to produce maximum emissions-reducing benefits. A battery electric vehicle can also be emissions-free, especially if the electricity it uses is from a clean grid. With the huge amount of funding being poured into powertrain research, largely driven by the globally converging fuel economy and greenhouse gas standards, the pace of innovation is rapidly accelerating. After a hundred years dominated by the internal combustion engine, this investment is resulting in an expanding range of cleaner alternatives.

## A Cleaner ICE

The internal combustion engine—known as an "ICE" in the industry—that is under the hood of most vehicles has been vilified and labeled an environmental problem since the early 1970s. The September 4, 1970, edition of *Life* magazine even suggested that the engine might be supplanted by electric or other cleaner forms of propulsion within ten years.[1] But, more than four decades later, the ICE is still the source of most greenhouse gas emissions and retains a near monopoly within the transportation sector: virtually 97 percent of vehicles are still powered by gasoline or diesel-burning internal combustion engines. Ironically, it is in that same internal combustion engine where some of the initial high-impact upgrades in cleaner, smarter technology are occurring and will continue to occur.

Since 2001, I've primarily driven hybrids or plug-in hybrids. I know many of my passionate environmentalist friends would love to see gas-burning cars completely disappear tomorrow. But the internal combustion engine will be a primary personal transport option for a while. Fortunately, the potential for further improving internal combustion powertrains is far from exhausted.

Part of the reason is that the vehicles are still so inefficient. The Department of Energy reports that a mere 14 to 26 percent of the energy in fuel tanks is actually used to move a current generation ICE vehicle down the road. While it is impossible for an engine to reach 100 percent thermal efficiency, huge gains will still be made in the near future. Toshihiro Mibe, Honda's managing officer of automotive development, says his company is "aiming at mean thermal efficiency of 50 percent."[2] Honda and other companies are already reducing heat-related losses, lowering friction in engines, and powering accessories like air conditioning electrically. They are also using and developing smaller turbocharged engines and more efficient transmissions.

It was these types of technologies, already in existence but underutilized, that were most important in convincing automakers to sign on to our 2017–2025 standards. Chet's team out in Ann Arbor worked extensively with the engineers from Ford, GM, and Chrysler to show them that our proposal didn't require some technical miracle or commercially unviable proposition. The car companies would have panicked if they thought we were going to require them to sell 40 percent electric vehicles by 2025 to meet the standards. We had to show that it was possible to meet the rules profitably with proven technologies. In fact, the 54.5–mpg fleet average is based on the assumption that only 1 to 3 percent of automobile sales in 2025 will have to be electric vehicles or plug-in electrics. The rest can be met with conventional internal combustion–powered vehicles and their hybridized versions.

Using the research we had done, we developed potential scenarios for each auto manufacturer, leveraging largely existing technologies. The Ann Arbor EPA team shared their analyses with their industry counterparts. In the specific case of Ford, for example, we calculated that the company could put high efficiency gearboxes in 97 percent of its vehicles, almost its entire fleet. The company could also have automatic transmission with eight forward gears in about half its cars and trucks.

We showed Honda that they could switch 10 percent of their fleet to what are called mild hybrids. Mild hybrids utilize stop-start technology, which effectively turns the engine off when a car is at a stoplight or stuck in traffic instead of idling. When the driver steps on the accelerator, the car turns on again seamlessly. The technology is offers a 5–10 percent fuel economy improvement and is affordable at around $300 per car.[116] Our rules assumed that as many as 45 percent of the total vehicles sold in 2025 would have some form of this technology. While about 40 percent of European cars already use the technology, American car companies are rapidly closing the gap. Ford, for example, plans to have the technology on 70 percent of its models in 2017.

We also suggested that all manufacturers could add lower rolling resistance tires, engine friction reduction, turbocharging, and engine downsizing. These are not scary, theoretical technologies that may not work in practice. They are proven, necessary steps toward much more efficient vehicles.

★   ★   ★

One sign that car companies have embraced cleaner technologies is the sub-brands they are launching, investing in, and heavily marketing. It is now not only horsepower they are advertising but better fuel economy. What Ford calls its EcoBoost option uses turbocharging and direct injection to make smaller, more efficient four-cylinder engines perform like six-cylinder engines. The company markets EcoBoost for both its high efficiency and power. The 1.0-liter EcoBoost, engine in the 2014 Ford Fiesta has an EPA-estimated fuel efficiency rating of 31 mpg in the city and 43 mpg on the highway. On the other end of the range, four-cylinder EcoBoost-enabled engines in some of Ford's larger and higher performance vehicles like the Mustang and the Explorer produce over 300 horsepower.

GM's Active Fuel Management system goes the other way, starting with larger engines but shutting down cylinders when they aren't needed. During highway cruising, for example, GM six- or eight-cylinder vehicles may only use four cylinders to reduce fuel consumption. According to the EPA, the 6.2-liter, 455-hp Chevy Corvette gets a 29-mpg highway rating. GM's Active Fuel Management is also a much cheaper option than Ford's EcoBoost system. Ford generally charges about $1,000 for the EcoBoost option, while GM's Active Fuel Management only costs about $50–100 per engine.[3]

Japanese companies, who have long used "eco-branding" as a strategy, are also rolling out their next generation branded powertrains. Honda's Earth Dreams line integrates turbocharging, combining it with direct injection for both gasoline and diesel configurations. The new internal combustion Honda Fit with a 1.3-liter Earth Dreams engine has the same fuel economy as the previous generation hybrid Fit.

Mazda's SkyActiv uses high-pressure fuel injection, high compression engines, and lightweight materials to improve efficiency while retaining a sporty image. The company's 2012 line has already met the 2016 CAFE standards. The 2012 Mazda3 with the optional new powertrain carries an EPA fuel economy rating of 28 mpg city and 40 mpg highway. Compared with the car's previous powertrain, these are improvements of 16 percent and 21 percent, respectively,

Toyota, the father of eco-branding, has a new engine line called the Toyota New Global Architecture. The automaker's 1.3-liter and 1.0-liter engines are engineered to boost fuel economy by at least 10 percent. Combined with stop-start technology, the company estimates a 15 percent to 30 percent gain in fuel economy. In spring of 2014, Toyota announced that they would be offering a new series of three- and four-cylinder engines in many of their models.

These branded lines are some of the meat and potatoes technologies already available today or in the near future. While they aren't as glitzy as other innovations, they will play a huge role in improving fuel efficiency while reducing carbon dioxide emissions both immediately and for years to come.

A 2013 National Academy of Sciences report projects the development of cleaner, more efficient gas-powered vehicles far into the future. The report estimates that by 2030, the old polluting internal combustion engine can be transformed into a cleaner, more efficient machine that can raise fuel economy to 65 mpg for an average passenger car and, by 2050, get up to 87 mpg, a rarified position by today's standards.[4]

But improving traditional powertrains is just the beginning. To help meet our greenhouse gas emissions goals, the traditional gasoline-burning internal combustion engine needs to be largely replaced by 2050. Today, cars that will replace those gas-burning machines are multiplying rapidly.

## The Transition to Alternative Powertrains

At the turn of the nineteenth to the twentieth century, electric cars were the top-selling road vehicles throughout the United States, slightly ahead of both steam- and gas-powered cars. Moving at slow speeds for short distances, pure electric vehicles were marketed as a woman's product because of their silence, smoothness, and ease of operation.

Early electrics were also used in some of the first automotive fleet cars. At the end of the nineteenth century, Pop Manufacturing Company, the first large-scale electric vehicle manufacturer in the United States, built the more than fifty taxis to serve New York City. The cabs' main competition was horse-drawn carts. One advantage for the clean, quiet electrics was that they didn't leave a trail of manure behind them, a big problem in a crowded urban setting.

But the invention of the electric starter in 1912 initiated their decline, and by the mid 1920s they were pushed to the side by internal combustion vehicles, which had the advantage of range. Ever since then it's been hard for them to get back on the road. The technology still existed, but its use was limited to niche vehicles like golf carts. Alternative powertrain vehicles all but disappeared.

It was a 1990s requirement by California's Air Resources Board that effectively kick-started the modern day development of clean powertrain technologies. The zero emissions vehicle (ZEV) regulation required that 2 percent of vehicles for sale in California in 1998, and 10 percent of vehicles by 2003, be zero emission vehicles such as hydrogen fuel cell and battery electric vehicles.

GM's EV-1, General Motors' response to the mandate—and the first electric vehicle made by a major American automaker in the modern era—was leased to Californians in the late 1990s. Between 1996 and 1999, GM produced more than 1,100 of the cars. Drivers loved them, but GM concluded that they would be unprofitable and canceled the program in 2002, in part to show that California's concept of a zero emission vehicle was not feasible. The company repossessed all the vehicles, refusing to sell them to any of their passionately devoted lessees. Instead, the cars were taken to the desert and crushed for scrap metal. The documentary *Who Killed the Electric Car* details the episode, pointing a finger at not just automaker, but the oil industry, the US government, competition from hydrogen fuel cell proponents, and the California Air Resources board itself for eventually killing the ZEV requirement in the face of lawsuits. In 2001, the California Air Resources Board updated the ZEV 10 percent requirement to allow a mix of hybrids, fuel cells, and ultra clean gasoline vehicles in addition to electrics. The 10 percent requirement continues today, with a target of 1.5 million zero emissions vehicle on California's roads by 2025.

At the federal level, it was President Clinton's 1993 Partnership for Next Generation Vehicles (PNGV) program that added to the momentum started by California. The effort was a government-financed initiative to work jointly with automakers to develop extremely efficient vehicles capable of 80 mpg by 2003. The program's initial budget, as reported by the Congressional Office of Technology Assessment, was to be about $500 million over ten years, with industry contributing about $200 million and the government portion spread across a number of agencies.

As EPA's representative, I saw the program as a potentially a good step toward reducing greenhouse gas emissions in the transportation sector. Vice President Al Gore was a big supporter of the effort, once hosting us at the vice president's residence to personally encourage us to make this partnership a success. He was incredibly sincere and committed to the effort. Unfortunately, the program had a major flaw: the car companies had no intention of seeing it succeed. For various reasons, including their belief that such fuel-efficient cars would not be profitable, the program limped along, at a total cost of a billion dollars, until it was terminated early in President George W. Bush's administration.

The irony was that the American program unwittingly gave birth to a highly successful alternative powertrain technology—in Japan. Because foreign companies were not allowed to participate in the PNGV initiative, Toyota's chairman Eiji Toyoda decided to create his own program to develop a global car for the twenty-first century. In

1997, the fruit of the program, the Toyota Prius, came to the Japanese market. Like the EV-1, it was a revolutionary design, and early customers embraced the extremely fuel-efficient car. Rather than crushing the vehicles for scrap, the company turned Prius into its flagship and used it to promote the Toyota brand as eco-conscious and innovative. To date, its hybrid electric technology has been incorporated into more than seven million models sold worldwide.

In August 2009, early in his administration, President Barack Obama announced the most recent federal push for alternative powertrains as part of the massive American Recovery and Reinvestment Act of 2009, promoted to pull the American economy out of its tailspin. The act provided $2.8 billion, an unprecedented investment in the manufacturing and deployment of electric vehicles, batteries, and components in the United States. "If we want to reduce our dependence on oil, put Americans back to work, and reassert our manufacturing sector as one of the greatest in the world, we must produce the advanced, efficient vehicles of the future," said the president in announcing the funding.

The majority of the nearly $3 billion went into electrification, with $1.5 billion toward the manufacturing and deployment of the next generation of US batteries, $500 million for the manufacture of electric-drive components, and $400 million for transportation electrification, including home charging equipment and public charging networks. A further $300 million was allocated for the Clean Cities program to displace petroleum with alternative fuels and infrastructure. Funds were also provided for owners to acquire chargers at a reduced rate, for installation of a nationwide public charging network, and for a $7,500 tax credit to buyers of electric cars. Numerous states including California provided additional incentives in the form of tax credits, effectively lowering the purchase price.

In large part because of these initiatives, there are at least seventy-six alternative powertrain vehicle models sold in the United States in 2015 (Figure 8). Most start with fuel economy ratings of 40 mpg or more, and a number of them have outstanding EPA fuel economy ratings above 90 miles per gallon equivalent (mpge). The new term mpge was developed by the EPA specifically for electric vehicles to allow comparison with gasoline vehicles. It is based on the fact that one gallon of gasoline is equal to 33.7 kilowatt hours (kwh) of electricity. So, the distance traveled by 33.7 kwh's of electricity is equivalent to the same distance traveled by a gallon of gasoline. The Nissan Leaf's EPA rating is 114 mpge, while the Ford Focus Electric's rating is 105 mpge. The Chevy Volt, despite its additional internal combustion powertrain, has

**Figure 8. The Alternative Powertrain Vehicle Models Available in the US Market for MY 2015**

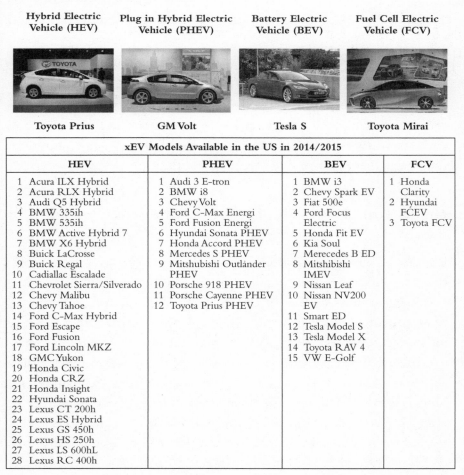

| Hybrid Electric Vehicle (HEV) | Plug in Hybrid Electric Vehicle (PHEV) | Battery Electric Vehicle (BEV) | Fuel Cell Electric Vehicle (FCV) |
|---|---|---|---|
| Toyota Prius | GM Volt | Tesla S | Toyota Mirai |

| xEV Models Available in the US in 2014/2015 | | | |
|---|---|---|---|
| **HEV** | **PHEV** | **BEV** | **FCV** |
| 1 Acura ILX Hybrid | 1 Audi 3 E-tron | 1 BMW i3 | 1 Honda Clarity |
| 2 Acura RLX Hybrid | 2 BMW i8 | 2 Chevy Spark EV | 2 Hyundai FCEV |
| 3 Audi Q5 Hybrid | 3 Chevy Volt | 3 Fiat 500e | 3 Toyota FCV |
| 4 BMW 335ih | 4 Ford C-Max Energi | 4 Ford Focus Electric | |
| 5 BMW 535ih | 5 Ford Fusion Energi | 5 Honda Fit EV | |
| 6 BMW Active Hybrid 7 | 6 Hyundai Sonata PHEV | 6 Kia Soul | |
| 7 BMW X6 Hybrid | 7 Honda Accord PHEV | 7 Merecedes B ED | |
| 8 Buick LaCrosse | 8 Mercedes S PHEV | 8 Mitshibishi IMEV | |
| 9 Buick Regal | 9 Mitshubishi Outlander PHEV | 9 Nissan Leaf | |
| 10 Cadiallac Escalade | 10 Porsche 918 PHEV | 10 Nissan NV200 EV | |
| 11 Chevrolet Sierra/Silverado | 11 Porsche Cayenne PHEV | 11 Smart ED | |
| 12 Chevy Malibu | 12 Toyota Prius PHEV | 12 Tesla Model S | |
| 13 Chevy Tahoe | | 13 Tesla Model X | |
| 14 Ford C-Max Hybrid | | 14 Toyota RAV 4 | |
| 15 Ford Escape | | 15 VW E-Golf | |
| 16 Ford Fusion | | | |
| 17 Ford Lincoln MKZ | | | |
| 18 GMC Yukon | | | |
| 19 Honda Civic | | | |
| 20 Honda CRZ | | | |
| 21 Honda Insight | | | |
| 22 Hyundai Sonata | | | |
| 23 Lexus CT 200h | | | |
| 24 Lexus ES Hybrid | | | |
| 25 Lexus GS 450h | | | |
| 26 Lexus HS 250h | | | |
| 27 Lexus LS 600hL | | | |
| 28 Lexus RC 400h | | | |

| xEV Models Available in the US in 2014/2015 | | | |
|---|---|---|---|
| **HEV** | **PHEV** | **BEV** | **FEV** |
| 29  Lexus RX 400h<br>30  Mazda Tribute<br>31  Mercedes ML450<br>32  Mercedes S400<br>33  Nissan Altima<br>34  Nissan Infiniti M35h<br>35  Nissan Infiniti Q50h<br>36  Porsche Cayenne<br>37  Porsche Panamera<br>38  Range Rover Hybrid<br>39  Toyota Avalon Hybrid<br>40  Toyota Camry<br>41  Toyota Higlander<br>42  Toyota Prius<br>43  Toyota Prius C<br>44  Toyota Prius V<br>45  VW Jetta Hybrid<br>46  VW Touareg Hybrid | | | |

a rating of 98 mpge when operating on electricity. The Tesla S is rated at an outstanding 95 mpge. The BMWi leads the pack at 117 mpge. Every one of these vehicles has benefitted from various sorts of government funding and incentives and will continue to be an essential part of meeting our 2050 greenhouse gas emissions goals, although their contributions will differ in scale and timing.

## Hybrid Electric Vehicle (HEV) = ICE + Electric Motor + Small Battery

The Prius, the first hybrid electric vehicle (HEV) was a "disruptor" that forever changed the market for alternative powertrain vehicles. Following its introduction in 1997, other major automakers began to offer at least one model with technology that went beyond the old internal combustion model. What followed was an incredibly rapid advance. About a decade ago there were only four hybrid models available in the US market; as of the 2015 model year there are 46 different models from 12 different automakers, both foreign and domestic.

An HEV has both a small internal combustion engine and an electric motor powered by a small battery. While driving, the battery is recharged primarily by capturing energy normally lost to the brakes, which in turn powers the electric motor. The electric motor runs in parallel with the gas-powered engine, providing a boost that reduces the load on the internal combustion engine, improving fuel efficiency. This allows most hybrids to achieve 40 to 50 mpg, and unlike internal combustion vehicles, their higher mileage ratings are in city driving.

Toyota continues to dominate the American hybrid electric market. In 2013, nearly half a million HEVs were sold in the United States, over 50 percent of which were Toyotas. Globally, Toyota also dominates the segment, selling 1.28 million hybrids globally in 2013 and a total of roughly 7 million to date. Toyota also remains the most ambitious. By 2020, the company plans to offer a hybrid configuration on every one of its models.

Today Toyota is far from the only company that sees hybrids as a wave of the future. The senior VP for BMW's i Product Line, Ulrich Kranz, told me. "We think that every car in 2030–2040 will have, in addition to a small ICE, some kind of an electric motor. Some will get energy out of Lithium ion or a next generation battery, some will get energy from hydrogen fuel cells"[5]

Other experts predict also that HEVs will have a very important role in reducing greenhouse gases in the near future as we transition away from gas and diesel engines. Navigant Research forecasts that global HEV sales will approach 3.5 million in 2020 for all brands, up from around 2M in 2014.[6] The National Academy of Sciences estimates that by 2030, an average hybrid electric passenger car will be averaging up to 78 mpg and by 2050, up to 112 mpg. But by mid-century it should be supplanted by even cleaner technology.[7]

### *Battery Electric Vehicle (BEV) = Electric Motor + Battery Pack (+ICE Optional)*

It won 2013 *Motor Trend* and *Automobile* magazine's 2013 Car of the Year, made *Time*'s list of the twenty-five best inventions in 2012, got a perfect 5.0 crash rating from NHTSA, and was *Consumer Reports'* top-scoring car ever. The revolutionary Tesla Model S, a battery electric vehicle (BEV), goes 0–60 in 3.9 seconds, has an EPA rating of 95 mpge, and can travel 265 miles on one charge.[8] But there remains one barrier that even the Tesla has so far been unable to break through: price. The Tesla model with the 265-mile range option costs over $100,000. For the 160-mile range base model, drivers will still pay around $70,000, still expensive but closer to a high-end luxury car price. The more modest Nissan Leaf, the best-selling BEV, also costs about $10,000 to $15,000 more than a gas-powered equivalent, and that gap is shrinking every year. Largely as a result, sales have been slow to scale up in both in the United States and globally. As of October 2014, when Nissan already surpassed its 2013 total year sales, it had sold more than 66,500 Leafs in the United States since 2011, while the Tesla S has sold about 30,000 units since its 2012 debut.

Both the Nissan Leaf and the Tesla have passionate leaders behind them, albeit very different ones. Carlos Ghosn is the decade-long chairman and CEO of the Renault Nissan Alliance behind the Nissan Leaf. For him, making the mass-market electric vehicle successful has become virtually a personal quest, and many would argue he has bet the fortunes of Nissan on doing so.

Tesla is the brainchild of Elon Musk, a man who breaks the norm of a traditional auto industry leader. He is a dreamer, an innovator, and somebody who sees beyond what is deemed possible. His quest has been to also commercialize the electric vehicle, though beginning with a high performance, luxury sports vehicle that could be mistaken for an exotic Italian model. I met him briefly in 2011 at the opening of the TESLA office in DC, where alongside the striking Tesla S he also was displaying his other love, the Space X module. He is an intense man, with a presence to fill the room. We talked briefly about electric cars and his commitment to provide battery cost data to our engineers. His passionate commitment to electric vehicles was clearly visible, and he may help change the world with his vision.

A battery electric vehicle is a much simpler design than a traditional internal combustion vehicle. It has an electric motor, which drives the wheels. There is no complex engine compartment, the transmission is one or two speeds only, and there is no gas tank with fuel lines running from back to front. The "fuel" is in the battery pack, and the battery pack is charged by plugging it in to either a regular 110V wall socket or a faster, higher voltage charger. Because they have no greenhouse gases coming out the tailpipe, fully electric cars have a huge potential for advances in greenhouse gas reduction. But they need cheaper, longer range, faster charging batteries—as well as an expanded infrastructure of charging stations—before they can become a larger part of the overall automobile market.

In the case of both the Nissan Leaf and the Tesla, the heavy battery makes up about half the cost. Both vehicles are powered by lithium-ion batteries, the same technology in iPhones, laptops, and all sorts of other gadgets. Though the power source revolutionized electronics, it is quite expensive and hard to scale up to power something as big as a car.

Additionally, most mass-market battery electric vehicles, like the Leaf, are limited to around a 100-mile range between recharges for the 24 kilowatt-hour (kwh) battery pack it comes with. A kilowatt hour is a common measurement of electrical energy; it's the unit most commonly used on utility bills to measure how much electricity a house or business uses. For electric vehicles, kwh is used to rate the electrical energy storage capacity of the batteries in the vehicle. Thus a vehicle with a larger capacity battery pack will have an extended range between charge-ups but cost more.

Recharging the batteries of a vehicle like the Leaf typically takes about four hours on a 240-volt charger and eight hours if plugged in to a regular 110-volt socket. This makes the cars fine for most urban driving, but impractical for extended highway driving. The interim solution has been a variant of the BEV, which adds an internal combustion powertrain to help address the short driving range of the battery. These are referred to as plug-in hybrids (PHEVs). For example, there is a plug-in version of the Prius that can travel purely on electric power for about ten miles before the internal combustion engine kicks back in and the vehicle reverts to a normal hybrid operation. The battery pack is smaller, and thus less expensive, but that is offset by the need to incorporate a traditional internal combustion power-train into the vehicle.

The Chevrolet Volt popularized PHEV technology to Americans, but GM prefers to call it an "extended range battery electric vehicle" because it can travel about three hundred miles without recharging or fueling. It has an electric-only range of around forty miles. The internal combustion engine onboard isn't directly connected to the wheels. Instead it acts like a generator, kicking in when the battery is drained and producing electricity to power the electric drive motor linked to the wheels. The newly announced BMW i3 BEV has adopted the same architecture, but because of its 100-mile battery-only range, it uses only a very small internal combustion engine to add about sixty miles of range to the car.

Because most trips Americans take are less than forty miles, the Volt doesn't use its gas engine often, making its fuel efficiency much higher than that of a traditional hybrid. The Volt gets a combined EPA city/highway mileage rating of 98 mpge when driven on battery electricity and 37 mpg when driving on gasoline-powered electricity, versus the Prius's 50 mpg. I have driven a Volt for the last three years, and, aside from visits to my daughter in Baltimore, all my daily trips are on electricity. Plugging in the car after each trip sounds unusual and maybe even intimidating, but it's no more complicated than closing and opening the car door.

Though the plug-in hybrid electric vehicle is relatively new technology, it's growing rapidly and offers some of the greatest diversity in price and configuration. The best news is that plug-in electric vehicle sales—PHEV plus BEV—are growing faster than historical hybrid sales. Sales of the combined category in 2013 exceeded 200,000 units, and there are more than 500,000 on the road now worldwide. The Nissan Leaf alone had sold more than 100,000 units worldwide as of the start of 2014, and the Volt has sold 65,000 units since the start of production late in 2010. As of 2015, there are some twelve plug-in hybrid electrics available in the United States, ranging from the $845,000 Porsche PHEV Spyder to the models from mass-market manufacturers like GM, Ford, Honda, and Hyundai, whose PHEVs are now in the $30–$35,000 range. If the exponential rate of growth continues, we would have some three million plug-in electrics on the road worldwide by 2020.[9]

An electric vehicle requires the consumer to think differently about price and cost of a car. We are conditioned to think about the sticker price of a car and not about the total cost of ownership when we make a purchase. With an electric car, the upfront sticker price is higher, but the operating costs are much lower. For example, Consumer Reports estimates that a Nissan Leaf costs about 3.5 cents per mile to operate on electricity, where a 2013 Toyota Corolla that with an EPA combined 32 mpg rating costs 12 cents per mile to operate.[10] The routine maintenance cost of EVs is generally also significantly lower—they don't need oil changes, and brake pads tend to last longer, for example. So over the life of the two cars, the total cost of operation will be different, depending how many miles they are driven. Electric vehicles will become more attractive to consumers once they adopt this new criterion when considering a purchase.

But the biggest single factor in increasing consumer adoption of EVs is how fast and how far battery costs come down, reducing both upfront costs and the total cost of ownership. Fortunately, they have halved in the past four years and are on track to drop further. The Electric Vehicles Initiative believes that battery pack costs could be reduced to around $300 per kilowatt hour by 2020.[11] If battery packs cost $300 per kilowatt hour in 2020, that means that electric vehicles will cost roughly the same as equivalent internal combustion vehicles on a life-cycle cost basis. In other words, a driver would pay a higher upfront cost to buy a BEV, but over the useful life of that vehicle the reduced operating costs will make it equivalent to the total cost of owning an equivalent gasoline-powered car or truck. Depending on how you interpret the data, it is arguable that Tesla has already achieved $250 or less per kwh for their battery packs. Tesla points out

**Figure 9. Total Cost of Ownership of Various Powertrain Technologies**

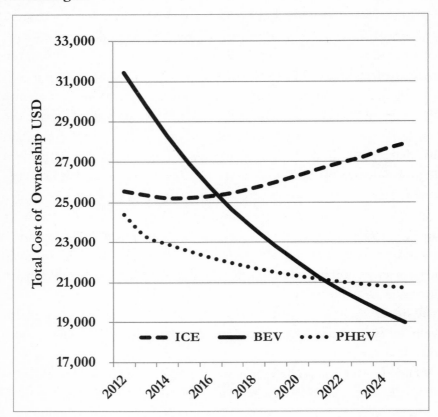

*(Source: Electrification Coalition, PRTM Management Consultants Analyses)*

that the "gigafactory" for batteries they announced in 2014 is targeting $100/kwh by 2020.

The Electrification Coalition, a consortium of twenty-five companies involved in various elements of battery electric vehicle production, also suggests $300/kwh as the "tipping point" for BEV adoption. A federal tax credit already lowers the total cost of ownership for BEVs and PHEVs by $7,500. As shown in Figure 9, the Electrification Coalition's analysis demonstrates that PHEVs already enjoy a lifecycle cost advantage over ICEs and forecasts that BEVs and internal combustion engines could achieve cost parity as early as 2017.

Both automakers and government agencies like the Department of Energy have ambitious targets for future innovation. For example, in 2013, the Energy Department's labs initiated a Batteries and Energy Storage HUB program that brings together the best talent in industry and government labs to develop by 2018 a battery that is one-fifth the cost of current batteries but has five times the energy density. Modeled after the scientific management characteristics of the Manhattan Project and AT&T's famous Bell Laboratories, HUBs are integrated research centers that combine basic and applied research with engineering to accelerate scientific discovery and address critical energy issues.

If successful, this HUB project will deliver a battery pack cost of around $150 per kwh or less, while the improved energy density would mean a driving range five times greater than today. When the battery industry is successful in achieving these levels of battery costs, electrics will become less expensive to own and operate than internal combustion engine vehicles while delivering comparable range between recharges.

★ ★ ★

Besides cost, the other obstacle to wide-scale adoption of battery-powered cars is psychological, what is known as "range anxiety." Most daily trips are less than half of the Nissan Leaf's 120-mile range. Nonetheless, drivers—even those who drive almost exclusively short distances in large urban areas—get nervous about the fact that they can't drive for three hours on the open highway without finding a place to charge up.

As the former NHTSA head David Strickland put it, drivers want to be "able to jump on I-80 and go across America. . . . Americans are all about freedom and choice." The idea that, "I may not want to drive to Cleveland today, but with the vehicle I own, I have that choice," is critical for many American drivers. Obviously, improving the range and charging time of batteries will help assuage some of this concern. But providing more charging stations is a good second option.

To encourage their customers to freely drive around the country, Tesla has built its own network of charging stations. As of August 2014, the company operated 105 stations, with heavy coverage from Vancouver, British Columbia, down to San Diego and the whole West Coast, with stations spaced across the country on interstates to encourage transcontinental trips, plus an expanding number of charging spots in Europe. The Tesla "supercharger" network is part of the cost of the purchase of a Tesla S with the larger 85 kwh battery. As part of the network, drivers can get a 65 percent recharge in thirty minutes at a very high 400 volts. Because a 100 percent charge up takes up to four hours, this partial charge is a much better option for drivers taking a short break on a longer drive. Other companies offer subscription-based services. The Car Charging Group, for example, has over 1,250 public charging sites that offer 240-volt charging in forty states and fourteen countries.

But even the concept of "plugging in" to charge may soon be eliminated. The city of Gumi, South Korea, is testing an electric bus that is charged wirelessly, through a process called inductive charging. On a much smaller scale, the same technology is already used in homes across America to recharge electric toothbrushes. Toyota has announced that the next generation plug-in Prius—probably available before 2020—will combine advanced technology such as higher power density solid-state batteries with wireless charging. Solid-state batteries have solid electrodes as in a conventional lithium ion battery, but in place of liquid paste electrolytes, there is a solid electrolyte that can store twice as much energy as the conventional lithium ion batteries. Nissan is also planning on using inductive charging on the next generation Leaf. Imagine parking your car over an in-floor charger and letting it charge itself.

★   ★   ★

The spread of various kinds of electric vehicles adds another dimension to measuring greenhouse gas emissions. Plug-in hybrid vehicles and battery electric vehicles have zero tailpipe emissions when they

## Figure 10. Electric Vehicle GHG Pollution Ratings by US Electric Grid Region

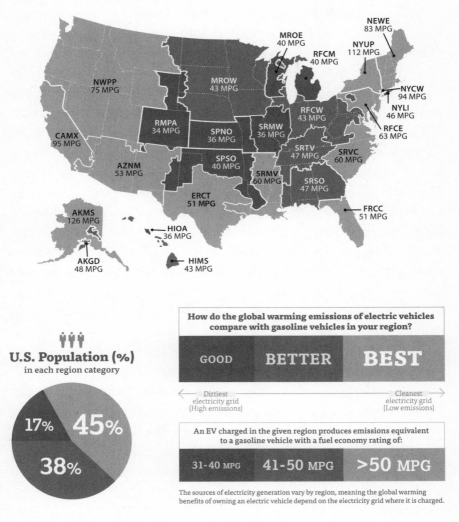

**U.S. Population (%)**
in each region category

**How do the global warming emissions of electric vehicles compare with gasoline vehicles in your region?**

| GOOD | BETTER | BEST |

← Dirtiest electricity grid (High emissions)          Cleanest electricity grid (Low emissions) →

An EV charged in the given region produces emissions equivalent to a gasoline vehicle with a fuel economy rating of:

| 31-40 MPG | 41-50 MPG | >50 MPG |

The sources of electricity generation vary by region, meaning the global warming benefits of owning an electric vehicle depend on the electricity grid where it is charged.

*(Source: Union of Concerned Scientists)*

run on electricity. However, the total carbon impact of driving a BEV or PHEV depends on how the electricity that they are charging with is produced. For example, a BEV powered by coal-generated electricity emits as much carbon dioxide into the atmosphere as driving a gasoline car rated around 30 mpg. The emissions inherent in a coal-generated electricity charge are called the "upstream" emissions; these are not presently measured in the EPA regulations.

Because different regions of the country produce electricity using different sources, geography can determine the total carbon footprint of identical cars. In the United States, this plays out along both state and energy grid lines. According to a 2012 Union of Concerned Scientists report, about 45 percent of Americans would produce fewer carbon emissions with battery electric than a 50-mpg gas engine (Figure 10).

In California, where my daughter Marisa lives, her Chevy Volt will get the equivalent of 70 mpg in carbon emissions due to increased use of gas, nuclear, and clean power there. However, because the Rocky Mountain grid covering Colorado and some neighboring states has the highest carbon intensity for power generation, a BEV charging up on that grid produces total carbon emissions comparable to that of a 34-mpg gas-powered Ford Fiesta. Fortunately, I live in a part of the country where the Chevrolet Volt I am driving at present does significantly better than a conventional car. When I drive it only on the battery, it gets the equivalent of about 58 mpg, 16 percent better than a Toyota Prius.

As we clean up greenhouse gases in the transportation sector, we must also reduce the carbon intensity of the nation's grid. President Obama has led the effort to ensure that electric vehicles and plug-in hybrid electrics achieve their full potential by asking EPA to regulate the carbon pollution from our nation's utilities. EPA has proposed a 30 percent reduction by 2030 from 2005 levels. Just as with the greenhouse gas rules my team worked on, this will require a concentrated effort.

★   ★   ★

According to the National Academy of Science, in order for battery electrics to contribute to the race against climate change, we need a combination of regulations, continued government investment and incentives, and private industry innovation to get them to an estimated 190 mpg by 2030 and 243 mpg in 2050.[12]

**Figure 11. Research, Development, and Demonstration (RD&D) by the EVI Countries**

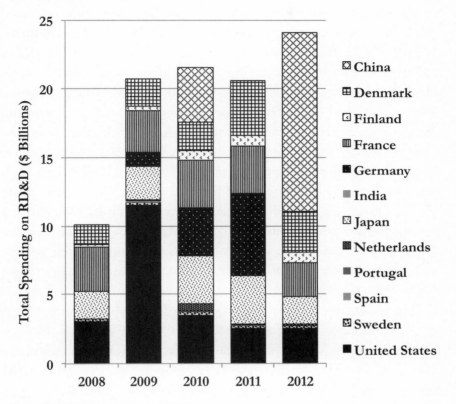

*(Source: EVI Global EV Outlook, April 2013)*

Meeting these goals will almost certainly depend on government funding of research and development. Fortunately, governments are making these expenditures, and fourteen have banded together in a coalition called the Electric Vehicle Initiative. The EVI coalition, which includes the United States, China, India, South Africa, and many European nations, among others, estimates that nearly $14 billion has already been spent since 2008 on research and development, fiscal incentives, and infrastructure investment to promote clean vehicle technologies.

Of that, $8 billion has gone into the research and development that has produced advances in alternative powertrains and battery technologies. Approximately $3.2 billion has been spent on fiscal incentives and another $1 billion on infrastructure.

Clearly, the United States was the early leader in investment, but France and Japan have been steady investors, as has Denmark (Figure 11). Looking to the future, however, China's government has announced plans to invest more on developing battery electric vehicle technology than the United States. With its "New Energy Vehicles Program," China hopes to provide its enormous numbers of future drivers with clean powertrains that do not further add to the emissions challenges the country already faces.[13]

But while all of these governments provided significant research and development investment and funding to help these "clean vehicles" gain traction, the different kinds of incentives used have affected the success of various national programs. The incentives are typically either direct subsidies—a one-time bonus upon purchase of vehicle or for charging equipment—or fiscal incentives like reduced purchase tax or reduction on annual tax.

The United States has relied mostly on direct subsidies in the form of a federal tax credit of $7,500 for a BEV/PHEV with a 16 kwh or larger battery. Smaller sizes earn proportional subsidies. Additionally, funds were provided for owners to acquire chargers at a reduced rate and for a public charging network to be installed nationwide. California and numerous other states have provided additional incentives in the form of state tax credits.

Some European countries have also offered direct subsidies to consumers, but these were typically linked to the carbon dioxide emissions of the vehicles. The tiny and fabulously wealthy nation of Monaco has offered the equivalent of $12,000 in subsidies for BEVs and PHEVs. France has offered up to $9,300 for vehicles that emit 20 grams of carbon dioxide per kilometer.

Most European countries have had more success in getting consumers into EVs by reducing or eliminating the relatively high European taxes on the vehicles. For example, Norway set a goal of putting

50,000 zero emission vehicles on the roads by 2018. It made BEVs largely tax exempt from most vehicle fees, including purchase taxes. Additionally, drivers of BEVs do not pay tolls. Taken together, these incentives can reach $11,000, making an electric car purchase price-competitive with conventional cars. In March 2014, Norway became the first country in which more than one in every 100 registered passenger cars is plug-in electric.

By comparison, Germany incentivized EV purchases but didn't relax sales taxes. Because BEVs and PHEVs have a higher initial cost than conventional versions of the same model, the sales taxes increase the cost of EVs disproportionately versus conventional models. This discrepancy may be part of the reason why Germany has had to back away from its 2009 goal of 1 million EVs on the roads by 2020. As of early 2014, there were only 4,600 EVs registered.

These lessons learned from the different countries' policies will hopefully translate into more successful programs in the years to come, until parity with internal combustion–powered vehicles is achieved both financially and in consumers' perceptions. By mid-century, electric vehicles and what many consider even a more a promising technology, fuel cells, must make up the majority of the new fleet.

★   ★   ★

### Fuel Cell Vehicles (FCEVs) = Electric Motor + Fuel Cell + Battery Pack

Fuel cells are such a futuristic-sounding clean technology that it's hard to believe models from Honda, Mercedes, and Hyundai have been on California's roads, albeit in limited volume, largely for continued testing and development. A fuel cell produces an electric current by forcing an electrochemical reaction between hydrogen and oxygen over a catalyst-coated membrane. Aside from electricity, the chemical reaction produces water. Proponents of the fuel cell like to point out that the technology is so clean you can drink the exhausted water from the tailpipe. In fact, the Apollo space missions used the cells for both electricity and drinking water.

Once the electricity is produced, the fuel cell operates like a battery in a battery electric vehicle and powers an electric motor. The current design of fuel cell vehicles integrates lithium-ion battery packs, like a BEV; the battery packs are recharged by the onboard fuel cell and used as a booster to help launch the vehicle.

The simple and clean design has made the fuel cell a sort of holy grail of green technology. Yet the slow adoption of fuel cells as a power source for electric vehicles has also made the technology a butt of jokes. In the auto industry, it has long been referred to as "ten years away for the last thirty years."

Fuel cell technology's slow advance has not been for a lack of interest or investment. Research for adapting the technology has come from both automakers and governments. GM has reportedly invested more than $2.5 billion, while Mercedes, Toyota, Hyundai, and Honda have spent billions. The Japanese government has heavily subsidized fuel cell development. California's Zero Emissions Vehicle mandate, the same rule that led to Chevy's EV-1, has also played a huge role in promoting the technology. Yet the challenge of making the fuel cell "stack"—the central unit where the electricity is produced—both durable for the life of a car and cost effective has proved very difficult.

Fortunately, the investments and R&D of the last several decades are starting to deliver results, and a fuel cell vehicle will be an affordable option for customers in the near future. In the late 1990s, when I had my first meeting with Honda in Japan, the company senior managers talked to me with a lot of pride about their efforts to develop the car of the future, the Honda fuel cell car. A few years later, Honda managers brought a prototype fuel cell car to DC for me to test drive, and they briefed me on their progress. I observed the evolution of the technology and its sharp reduction in cost. I was told that the car that had cost $1 million in the early 2000s would be around $50,000 to

$70,000 by 2015. Recent developments suggest those forecasts were largely accurate.

The year 2013 saw announcements from the world's leading auto manufacturers. Four fuel cell vehicle development partnerships were formed: Toyota/BMW, Honda/GM, Daimler/Ford/Nissan and Volkswagen/Ballard.[14] Honda, Toyota, and Hyundai announced their commitment to production vehicles in the near future, and Toyota actually launched the world's first production fuel cell car, the Toyota Mirai in 2014 in Japan, with California sales planned for mid-2015. The Mirai's performance is very similar to cars consumers are familiar with—it has a range of 300-plus miles on a five-minute fill-up, has a fuel cell with a rating of 114 kilowatts (153 horsepower), goes from 0 to 60 mph in 9 seconds, and is estimated at approximately 60 mpge. Understandably, it costs somewhat more at 6.7 million yen ($57,460) before taxes in Japan.[15] However, a subsidy of 2.02 million yen on FCV purchases will make the cost closer to $40,000, which is comparable to the plug-in Chevrolet Volt's price in 2011, when it was introduced. It is expected that the Mirai will be priced similarly at around $57,000 in California, but the combined US federal creditof $4000, plus the $5000 California fuel cell vehicle credit will reduce the price by only half as much as in Japan. By comparison, Hyundai's Tucson fuel cell vehicle is priced at 150 million won ($137,000) in the Korean market, with a subsidy of 60 million won that reduces the price to around $82,000.

The Department of Energy estimates that performance metrics such as power density, range refueling time, and cold temperature operations (e.g. below 0 degrees Fahrenheit/-18 degrees Celsius) are sufficient for consumer vehicles. Cost is also coming down, and DOE estimates that currently a 80 kilowatt fuel cell for a vehicle application costs approximately $55/kw. They believe the target cost of $40/kw is realistic for the 2020 time frame and suggest that the ultimate target of $30/kw will help fuel cell vehicles reach cost parity with internal combustion vehicles.[16]

Toyota Chairman Takeshi Uchiyamada, the father of the first-generation Prius, said he expected global sales to rise to "tens of thousands" in the 2020s.[17] Toyota plans to sell around four hundred in Japan and three hundred in the rest of the world in the first year.[18] A key reason for the slower ramp-up is infrastructure—that is, the availability of hydrogen fueling stations. The problem is even more acute than for electric vehicles, since, in the case of fuel cells unlike electrics, there are currently no systems for recharging your car with hydrogen at home. Building a fuel cell station costs about the same as a regular gas station—between $1 to $2 million—but without the potential customer base for the investment to pay back, requiring continuing support of governments and private investors.

And, even though the chemical reaction of fuel cells is much cleaner than burning gasoline, the vehicles do still need fuel from processes that are generally not emissions-free. Most hydrogen is produced centrally by combining natural gas with superheated steam and a catalyst. Once the hydrogen is produced, it is trucked to fueling stations as a liquid or compressed gas. Because of the natural gas involved, producing hydrogen this way also creates greenhouse gases.

There are two alternative processes for producing hydrogen. In one, a current is passed through water to split the hydrogen and oxygen molecules. This electrolysis method is always done at the fueling station but requires a lot of electricity, making it expensive, and the total carbon emissions depend on how the electricity is produced. Hydrogen can also be made from biogas and wastewater, but this technique is currently only practical for stationary power generation.

The DOE estimates today that hydrogen production costs are around $4 to $9 per gallon gasoline equivalent (GGE = $H_2$ kg x 1.019), and the target is to be at or below $4 per gallon. Industrial gas companies have stated publicly that today they can deliver by tube trailer hydrogen to a typical 150–200 kg/day station at $8/kg, and could deliver it for as low as $6/kg by pipeline.[19]

As a result, governments in California, Japan, Korea, and Germany are all building stations and developing further plans to support the technology. According to H2stations.org, there are 186 hydrogen refueling stations in operation worldwide as of March 2014. California has the most, with eighteen stations, and is planning to spend $2 billion to build approximately one hundred stations by 2023. Japan leads in Asia with thirty-one stations and plans to build as many new stations as California. Once fuel cell vehicles get to a station, they do have one advantage over electric vehicles: they can fill up on hydrogen in about the same time as a conventional car fills its tank with gas.

As fuel cell cars overcome these challenges, they will likely play an important role in the fleets of the future. Long-term fuel cell stack durability challenges still must be overcome, and the costs have to continue their downward slope, as with any nascent technology. The fact that fuel cells have commercial applications in stationary back-up power and the like is helping the technology to receive additional government funding. The DOE is providing some $172 million in funding for fiscal year 2014, and Japan's 2014 government budget is some $72 million. California alone is spending more than $20 million per year. The National Academy of Sciences estimate that fuel cell cars could achieve 122 mpge by 2030 and 166 mpge by 2050 and play a significant role in the cleaner cars of the future.[20]

★   ★   ★

The move from vehicles powered by hyper-efficient internal combustion engines will be one of the most noticed transformations over the next few decades. After more than a century of near monopoly, vehicles powered entirely by gas or diesel will need to surrender the roadways to cleaner, alternative vehicles in just a few decades. To meet a 180-mpg average for new cars in 2050, virtually all new vehicles sold in the United States should be a mix of plug-in hybrids, battery electric, and fuel cell vehicles. Any self-standing internal combustion vehicles or ICEs in hybrid powertrains would have to run on low carbon fuels.

But this sea change in what makes vehicles go will still stop short of the reductions we need. Fortunately, there are other innovations that can help move the transportation sector closer to our greenhouse gas emissions goals. Improvements in aerodynamics, tire rolling resistance, lubricants, and especially in lightweight materials can contribute greatly to transforming automakers' efforts to reduce greenhouse gases—and do so regardless of a vehicle's powertrain.

# 16

# LIGHTER AND STRONGER

From the day the first fuel economy standards went into effect, auto manufacturers have targeted vehicle weight. The physics is simple: it takes less energy, and thus fuel, to move a lighter vehicle down the road. The principle holds for all types of vehicles. A lighter internal combustion car will require a smaller, more fuel-efficient engine. A lighter battery electric can run on a smaller, cheaper battery while delivering the same range. According to the Department of Energy, reducing a car's mass by 10 percent can raise fuel efficiency numbers by 6 to 8 percent.

In the 1970s and '80s, automakers pursued these design goals before computing technology became pervasive. So, instead of running simulations, the companies hired engineers whose sole task was to painstakingly add up the weight of each of the thousands of parts of a proposed vehicle and then calculate its total "curb weight." On that basis, the company would select an engine that would help the automaker meet the overall CAFE targets. It was a thankless task that often led to failure. Engineers frequently overran the weight target estimates at the component level, creating a heavier-than-anticipated vehicle. The heftier car then required a larger, heavier engine, further lowering the fuel economy of the vehicle. A vicious circle of costly reengineering would begin in order to reduce the weight of the final vehicle to meet the fuel economy targets.

Nonetheless, average vehicle weights dropped dramatically as a result of the CAFE fuel efficiency standards. In 1975, the average car or pickup truck weighed about 4,200 pounds. By 1980, on a diet of strict mileage requirements, the fleet had downsized and dropped 1,000 pounds per vehicle, a 20 percent improvement. These same cars and trucks became 45 percent more fuel efficient at the same time, going from 13.1 mpg in 1975 to 19.2 mpg in 1980. Unfortunately, neither of these positive trends lasted long.

As shown in Figure 12, cars and trucks managed to keep their weight off until the mid-1980s, when fuel economy standards were frozen. Then, as SUVs and minivans began dominating the market in the 1990s, the weight of vehicles began edging upward. By 2010,

**Figure 12. Fuel Economy, Weight, and Horsepower for MY 1975–2013 (Adjusted)**

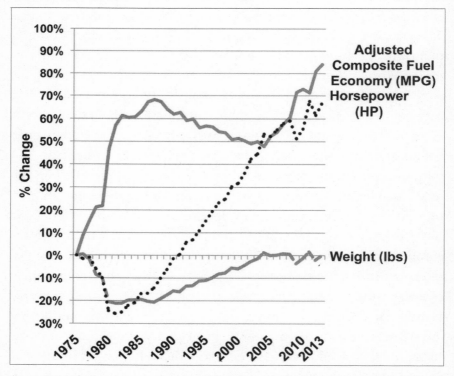

(*Source: EPA Light-Duty Automotive Technology, CO2 Emissions, and Fuel Economy Trends: 1975–2013*)

cars and trucks were back to the average weight of a vehicle in 1975. Likewise, fuel economy improved only marginally between 1985 and 2010. There had been lots of innovation in the previous thirty years. Unfortunately, all the research and development had gone into making heavier, much more powerful cars and trucks, and this was the reason for the stalled fuel economy numbers. Between 1980 and 2010, average horsepower increased nearly 70 percent.

In January of 2008, I stood in front of an audience of auto industry insiders at the annual Detroit Auto Show and asked them to rethink this obsession with horsepower. "We must bring about an end to the horsepower arms race among automakers and replace it with another, different kind of a race, a race to produce the most affordable and desirable, low carbon-vehicle each year."

The huge event has been held downtown in the Cobo Hall Convention Center every year since 1965. One of the most important exhibitions of new cars and truck models in the world, the Detroit Auto Show consistently featured revolutionary concept cars along with production models loaded up with fuel-slurping horsepower. That year was no exception. GM showed off a prototype of the Cadillac Provoq, an emissions-free electric car that used hydrogen fuel-cell technology to extend its range. Exciting, but the new Cadillac that people could actually buy was the CTS-V, a sports car with a 550 horsepower V-8 gas-powered engine. Other American companies offered new models with massive power. Dodge came to the table with a 600 horsepower Viper, while GM's new Chevy Corvette ZR-1 boasted a 638 horsepower engine. Needless to say, none of these cars were fuel economy leaders. The 2009 Corvette, for example, got 14 mpg in the city—worse than a century-old Model T.

For me, this horsepower race was so frustrating because I knew car companies could do so much better in reducing the carbon footprint of vehicles. The auto industry has always been full of innovation. If those resources were directed at the real problems facing us now, they could make a huge difference. An MIT economist named Christopher Knittel calculated that if automakers married the horsepower

and weight of cars at early 1980s levels with today's drivetrain technology, we would already have an average fleet efficiency of 37 mpg,
1.5 mpg better than our 2016 goals.[1]

With the new greenhouse gas and fuel efficiency standards set
through 2025, we finally have market innovation pointed in the right
direction. Instead of just producing heavier, more powerful vehicles
every year, automakers will be looking to lightweight materials to
improve the fuel efficiency and reduce the greenhouse gas emissions
of all kinds of vehicles, from traditional gas-powered internal combustion engines to trucks running on natural gas to pure electrics.

★   ★   ★

In designing our 2025 standards, we assumed that car companies
would meet the goals in part by reducing the weight of their vehicles 10 percent, on average. We also believed that companies could
go beyond 10 percent weight reduction without any safety issues. A
study done by Lotus engineering for the state of California had suggested that as much as a 30 percent reduction in weight was possible
without jeopardizing safety.

The Department of Energy has a program that is exploring the
manufacturing processes and costs for weight reduction in body powertrain, interior, chassis, suspension, glass, and joining technologies to
demonstrate fuel economy improvements from lower weight vehicles.
Department of Energy researcher Venkat Srinivasan says they want to
demonstrate a 50 percent weight reduction in passenger vehicles that
meet three additional criteria. The lighter vehicles must be just as safe
to drive, the improvements must be made without increasing cost, and
the program must be completed by the end of 2015.[2]

Our partner, the National Highway Traffic Safety Administration,
has pushed the limits of cost-effective, safe weight reduction as well. A
team of engineers took apart a 4,000-pound 2011 Honda Accord and
then rebuilt it using lightweight components, reducing the weight by
700 pounds. Their goal was to create a vehicle within 10 percent of
the original vehicle cost and with no compromise to safety. They not

only succeeded but beat their target, producing a lighter vehicle that achieved the same performance on a lower cost 1.8 liter 140 horse-power engine than the heavier and less fuel-efficient 2.4 liter 177 horsepower factory-installed engine.

Many automotive product planners have also set internal goals for reducing vehicle weight or, as the industry calls it, lightweighting. Carla Baido, Nissan America senior vice president for research and development, has said that next-generation vehicles must weight as much as 30 percent less to meet future fuel economy standards. And automakers must make the required improvements to their cars and trucks without additional costs. Fortunately, many more options are available to engineers than in the 1970s, when they had to individually weigh nuts and bolts.

**Figure 13. Change in Automotive Material Mix 1977–2035**

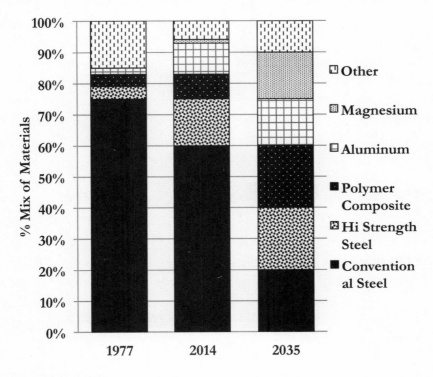

(Source: DOE, Automotive News)

★   ★   ★

To meet such ambitious goals, engineers have had to rethink how cars and trucks are made. As shown in Figure 13, in 1977 roughly 75 percent of the materials in a typical car were conventional carbon steel. As of 2014, high strength steel has increased to 15 percent and aluminum to around 10 percent, both displacing conventional carbon steel. By 2035, the Department of Energy suggests, cars will be comprised of only 20 percent conventional steel. Another 20 percent could be made up of high-strength steel—which can be 20 to 30 percent lighter than carbon steel, with the same strength—along with aluminum, composites, and magnesium.

**Figure 14. Comparison of Various Automotive Materials' Weight and Costs**

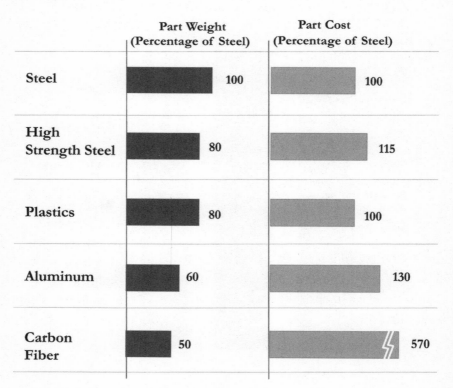

| | Part Weight (Percentage of Steel) | Part Cost (Percentage of Steel) |
|---|---|---|
| Steel | 100 | 100 |
| High Strength Steel | 80 | 115 |
| Plastics | 80 | 100 |
| Aluminum | 60 | 130 |
| Carbon Fiber | 50 | 570 |

(Source: McKinsey; Assuming 60K pieces/year)

In the hyper cost-conscious auto industry, materials selection and substitution is challenging. Each material offers trade-offs between weight and strength. Figure 14 offers a comparison done by the global consulting firm McKinsey, showing the relative weight and cost of different materials used in an automotive fender application.[3] When weight reduction is not a consideration, the higher cost materials become just that—added cost. But when weight reduction using more expensive materials produces a fuel economy advantage, then the higher cost of the lightweight materials becomes easier to justify.

As shown in the graph, high strength steel costs about 15 percent more than conventional steel but still doesn't deliver the same weight reduction as aluminum. Aluminum, however, costs about three times more than conventional steel for the same weight. But aluminum weighs less, so, on a weight-equalized basis, the cost penalty for aluminum versus high-strength steel can be offset in some applications.

As early as 2017, nanotechnology may be used create a new generation of thinner and stronger high-strength steels. Cars are likely to incorporate materials like aluminum and magnesium, as well as some materials that haven't yet realized their potential, including carbon fiber composites. The body engineer's job, historically dominated by working with conventional steel, has grown that much more complex, and perhaps exciting.

★   ★   ★

Aluminum's increasing popularity is easy to explain. The alloy is lightweight and high in strength as well as corrosion-resistant. Unfortunately, it is also energy-intensive to produce, making it more expensive than steel. Because of the expense associated with the material, aluminum has historically been used disproportionately by luxury automakers, like Audi. But because such companies have worked with aluminum for years, they have solved many of the initial manufacturing challenges. Aluminum is now found on many less expensive mass-market vehicles. The 2013 Honda Accord features an aluminum hood sub-frame and rear bumper. The 2014 Chevrolet Silverado and 2014

GMC Sierra both have aluminum hoods and suspension. The North American car of the year for 2013, Cadillac ATS, has an aluminum hood and many other aluminum components. Nonetheless, Ford is perhaps the most aggressive adapter of aluminum in the construction of mass-market vehicles.

In 2010, after the first greenhouse gas standard for 2012–2016 was finalized, Chet France, his engineering team, and I all visited Ford's Sue Cischke at the Ford headquarters in Dearborn, Michigan. We were there to feel out Ford's capabilities as we contemplated the next round of standards. Sue invited Chet and me into her office.

"We've been working on using more aluminum in our truck models—and perhaps extending it to other vehicles."

We were interested and asked her some questions until Alan Mullaly, Ford's CEO, walked in to greet us and thank us for collaborating with Ford. I knew Mullaly had been an engineer and executive at Boeing. The aircraft sector has been a pioneer in adopting lightweight materials like aluminum in their products to improve fuel economy. Perhaps Mullaly's familiarity with aluminum was part of Ford's rapid adoption of the metal.

At any rate, Sue was right: Ford went all in with aluminum. About four years later, at the 2014 Detroit auto show, the company showed off the all-aluminum body panel design for the new F-150. The company couldn't have made a stronger statement about its commitment to alternative materials. The F-150 is Ford's flagship vehicle, and the company sells about 650,000 of the trucks each year. With an Ecoboost engine, Ford expects the 700-pound slimmer aluminum F-150 jump in fuel efficiency from 23 to 30 mpg. By 2017, Ford plans to make more than one million vehicles with aluminum bodies.

Manufacturing experts and steel industry advocates say that moving to aluminum will require fundamental changes in how bodies are made all the way down the assembly line. It will also change the way the body is assembled, relying a lot more on adhesives and rivets then on welding alone.

Wolfgang Ziebart, Jaguar Land Rover's group engineering director, observed that, "the hard transition is when the decision is made to make the whole structure from aluminum."[4] The difficulty comes from both making the necessary improvements in the joining technology used in manufacturing and sourcing the aluminum itself, which is expensive. In the new Range Rover, which relies heavily on aluminum, Land Rover believes they have saved 880 pounds per vehicle. Each Range Rover body has 403 separate components, including stampings, and cast pieces that are held together with industrial adhesive and rivets. The old steel Range Rover that was discontinued in 2012 used more than 6,000 spot welds, and almost double the number of rivets that hold the aluminum Range Rover body together.

The competition between steel and aluminum will continue for a while longer. Long a champion of increased aluminum use in its more expensive Audi line, VW has decided to trade aluminum for high-strength steel for the new Golf, which is up to six times stronger than if made with conventional steel. This move helped VW reduce costs and about 100 kg in weight.

★   ★   ★

Mass-market car companies are working to introduce another lightweight construction material besides aluminum from an even more expensive and rarified segment of the automotive market. In the last decade, carbon fiber has moved from the racetrack, where it has been a staple for decades, into road supercars like the McLaren and Lamborghini. The material costs seventy-five to a hundred times as much as steel but is much lighter and stronger per square inch. The wasp-shaped McLaren P-1 breaks down into front, central, and rear body sections. Its carbon fiber construction is so light that an adult can lift the entire rear section with one hand, albeit clumsily.

Though car companies like BMW would love to introduce such an advanced material into their vehicles, carbon fiber is currently so expensive to manufacture that it has proven cost-effective for only very few industries, among them aviation. In part because of limited demand, six companies produce nearly 93 percent of the world's supply. As of 2012, production was limited to a capacity about 150 to 160 million net pounds of carbon fiber globally. But as Venkat Srinivasan from the Department of Energy's Lawrence Berkeley National Lab told me, "Most of it goes to aerospace and defense, a very profitable and protected market. If every vehicle made in the US used ten pounds of carbon fiber, the world's current supply would be depleted."[5]

Building additional plants to increase supply takes considerable resources. Today, it takes at least $100 million and as much as three years to build a production facility that can make five to eight million pounds of carbon fiber annually. Charles David Warren of the Department of Energy's Oakridge Labs estimates that, to be viable in the auto industry, the cost of manufactured carbon fiber has to come down by more than 50 percent from its current price of around $10 a pound. Such a cost reduction will likely require cheaper, lower performance carbon fiber. The creation of lower automotive grade versus aerospace grade carbon fiber is already being explored. The online magazine *Composites World* estimates that global net capacity will increase by 2020 to about 250 million net pounds, a 100 million–pound increase from 2012 levels.[6] But if automotive use does ramp up faster than assumed, there will be a shortage of supply.

Despite supply issues, carbon fiber is gradually becoming more prevalent in consumer vehicles. To reduce the enormous labor costs in producing the material, BMW has developed a mass production technique that uses a three stage semi-automated process. The German carmaker currently makes the passenger compartment of its i3 and i8 concept electric vehicles entirely out of carbon fiber–reinforced plastic and aluminum, much like the Boeing 787 Dreamliner. The lightweight i3, with around 660 pounds of carbon fiber content, weights a total of 2,700 pounds, including the 450 pound battery pack. A comparably sized BMW with a gasoline powertrain weighs about 700 to 800 pounds more.

Ford too has been experimenting with aluminum and composite materials, in collaboration with the US Department of Energy's Vehicle Technologies Program. The company has built a prototype Ford

Fusion weighing in at 2,600 pounds, about 800 pounds lighter than the 2013 base model and comparable to the super-mini Ford Fiesta. The prototype Fusion is a great example of how multiple different lightweighting techniques will converge in a single vehicle. Most of the structure is aluminum, with some magnesium in the transmission. One third of the Fusion's weight reduction is from advances in the interior and the glass, which is chemically laminated like on a cell phone. The front springs are composite, while the dashboard, oil pan, wheels, and seats are all carbon fiber. It is not a cost-effective design for production yet, but it does demonstrate the weight reductions possible from an integrated collection of advanced materials.

As futuristic as the Fusion's construction sounds, one of Ford's other prototypes goes back to the future. From the nineteenth century into the early twentieth, many cars were wooden carriages with engines added. Now Ford is taking the manufacture of composite materials back to the forest. In collaboration with the paper company Weyerhaeuser and Johnson Controls, Ford is experimenting with producing cellulose fibers from sustainably grown trees instead of fiberglass or mineral reinforcements. The company's research suggests that the plastic composite made from forest-grown cellulose meet its stringent durability requirements yet weighs 10 percent less than alternatives. The 2014 Lincoln MKX features the use of Cellulose Reinforced Polypropylene (called "THRIVE" composites by Weyerhaeuser) in a structural piece located within the center console armrest.[7]

Whether or not tree fibers have an important role in future cars, auto designers and engineers will be working with an increasing array of materials with a wide range of costs, strengths, weights, and other factors. The transformation produced by this convergence will be so far-reaching that before 2050 the construction of cars may resemble that of airplanes more than of any cars on the road today. The Boeing 787 Dreamliner is primarily carbon fiber but also contains substantial amounts of steel and aluminum fused together in ways that optimize each material in different capacities.

★  ★  ★

This change in materials will continue the century-long trend away from the boxy design of the original Model-T's, but it may also spell the end of Henry Ford's assembly lines. For example, 3-D printing allows products to be produced on demand and with much less waste than traditional manufacturing. Also known as Additive Layer Manufacturing, the technology literally grows materials particle by particle, like a bee making a honeycomb. As Andrew Brown, vice president and chief technologist of the Delphi Corporation explained, "3D manufacturing would reduce the need for large manufacturing facilities. So, instead of producing hundreds of thousands of something, you can produce a few of them and do it very cost competitively through additive layer manufacturing.[8]

As with the Boeing 787, this process is already in use in the airline industry. In 2014, the new Airbus 350 took flight with brackets created through 3-D printing. Whether or not 3–D printing is a central part of the process, there will definitely be commercially available super-light vehicle bodies manufactured or organically grown as an integral, single unit. Customers will potentially customize the products before they are "printed" or "grown."

Such innovations in creating low emissions vehicles—from relatively basic ideas like more efficient gearboxes to exotic cars built out of trees—are all in keeping with EPA's regulatory history of driving technological innovation. Just as with the catalytic converter in the 1970s, setting stringent standards generates stronger market demand for, say, fuel cells or stop-start technology. But, even with the breakthroughs on hybrids, EVs, and lightweight materials that we can expect in the coming decades, we will still be short. No matter how I run the numbers, I can't imagine a scenario in which we meet the 2050 greenhouse gas emissions goals for transportation unless we adopt additional policies, including the use of low carbon fuels in cars that continue to use an ICE powertrain.

# DEFOSSILIZING FUELS

In a sense, internal combustion vehicles don't create any greenhouse gas emissions. It's the fossilized carbon in the gasoline or diesel, when burned, that feeds global warming. All the other technologies designed to make internal combustion cars lighter and more fuel-efficient are simply trying to use less of that petroleum-based fuel. This means that the successful development and adaptation of transportation technologies that use renewable biofuels, electricity, and hydrogen fuel cells will be essential to reducing greenhouse gas emissions by 80 percent by 2050. But not all alternative fuels produce the same greenhouse gas reductions.

Honda was the first to offer a natural gas Civic, dating back to 2009, but it has more competition now. America's newly discovered stores of cheap natural gas have prompted new additions, including the Dodge Ram, and the 2015 Chevy Impala, to adapt their engines to run on the fuel. Today, the cost differential of natural gas versus gasoline or diesel is a major factor in some economists' and commodities experts' predictions of a decline in world oil use. A website that provides natural gas prices and filling stations nationwide, cngprices. com, lists the average summer 2014 price for compressed natural gas (CNG) at $2.23 a gallon, with some local markets going as low as 60 cents per gallon.[1] The United States hasn't had average gasoline prices that low since the end of 2009, when the country was beginning to dig its way out of a massive recession.[2]

In part because finding natural gas filling stations is difficult, the Honda Civic is currently the only dedicated CNG passenger car offered in the United States. Both GM's 2015 Chevy Impala and Chrysler's Ram Truck have engines that can switch from petroleum-based fuel to natural gas with no interruption, allowing drivers have the convenience of filling up at any gas station while also using a cheaper, cleaner fuel when they want. The Ram is designed primarily for truck fleets with central refueling stations.

While the low cost of natural gas immediately makes it attractive, its impact on carbon emissions is moderate. When powering a vehicle, the fuel produces 20 percent fewer greenhouse gas emissions, a significant but nowhere near sufficient improvement in reducing the buildup of carbon dioxide in the atmosphere. What's more, the Impala gets less than 20 mpg, hardly the prototype for a cleaner, more efficient future. Besides that, the production and distribution process used to extract and transport natural gas frequently releases large amounts of methane. Though it exists in relatively low atmospheric concentrations, methane is a greenhouse gas that is thirty times more potent than carbon dioxide. As a result, transportation uses of natural gas may actually be worse in terms of global warming than diesel or gasoline.[3]

Natural gas can still be useful as a short-term bridge between gasoline and low carbon alternative fuels. In addition, it could be the source material for compressed hydrogen used in fuel cell vehicles. However, the most important role for natural gas would be in the utility sector. Natural gas is the most likely energy source to replace coal at power plants, helping them meet the new EPA regulations requiring a 30 percent reduction of carbon pollution by 2030. However, the net impact on transportation would be positive. Not only would power plants reduce their carbon emissions, but the reductions from the energy sector could in turn create lower carbon electricity that can be used to power electric vehicles as well as plug-in hybrid electrics.

In 2013, the US Energy Information Agency predicted that natural gas will represent about 6 percent of transportation fuel by 2040, up from almost zero today. But there is another alternative fuel with a much better impact on carbon emissions.

★   ★   ★

While many nations adjusted their energy policies in the aftermath of the OPEC oil embargo, Brazil set the bar for bold conversion to biofuels. In 1975, the year the United States passed legislation to promote more efficient use of oil, Brazil began a program to completely eliminate petroleum-based fuels in its nation's cars. A key reason is that Brazilian sugarcane is relatively cheap, and production of biofuels from it is very energy efficient. The country had previously produced cars that ran on ethanol blends, and Brazil rapidly moved toward widespread usage of a purely sugar cane-based fuel.

In 1979 the country produced the world's first modern vehicle that ran entirely on ethanol. Even as oil prices plunged in the 1980s, Brazil maintained minimum ethanol blend requirements. Today, all fuel sold in the country contains at least 25 percent ethanol. Nearly all cars sold in Brazil, 94 percent, are flex-fuel, meaning they can run on any mixture of ethanol and gasoline. By 2018, Sao Paolo, a megacity of over 20 million, will be running all its city buses on renewable fuels, primarily ethanol.[4] Such a complete transition away from oil dependency is inspiring and unparalleled in any other nation. The United States also uses an increasing amount of biofuel—it is, in fact, the world's largest producer of ethanol—but it has followed a very different, politically complicated route.

In the 1970s and 1980s, the United States also used ethanol, but the percentage—around 2 to 3 percent—was much smaller than Brazil's. As lead was removed from gasoline, the oil industry looked at ethanol as an additive to offset the loss in octane. Adding about 2 percent ethanol to gasoline can also help reduce harmful tailpipe emissions and displace the use of toxic gasoline components such as benzene, a carcinogen. In 1990, the amendments to the Clean Air Act required the use of reformulated gasoline with either ethanol or methyl tertiary-butyl ether in areas of the country with serious air pollution. Nonetheless, corn ethanol usage didn't really grow until President Bush signed the 2005 Energy Policy Act.

As an initial move toward biofuels, the new energy law could be seen as exciting. The new legislation required that by 2006, 4 billion gallons, and by 2012, 7.5 billion gallons of renewable fuels be blended with the gasoline used in the transportation sector. To put that into perspective, close to 135 billion gallons of fuel were used in 2012 for transportation in the United States, so this meant that about 5.5 percent of that fuel would be from renewable sources.

But this novel measure was not driven by environmental concerns about the greenhouse gas emissions of petroleum-based fuels. In fact, the act provided billions in tax credits to fossil fuel companies and what the *Washington Post* had called "needless subsidies for oil drilling" in a July 28, 2005, editorial. The law also had the strong support of the congressional delegations from the Midwestern and Plains states, where most corn is grown.

The 2005 energy law also amended the Clean Air Act to give EPA the authority to implement this program. I called up Chet at the Ann Arbor EPA lab and asked him to put together a team that could manage the program. Because we had broad support from all the stakeholders, the process was pretty straightforward. The farm lobby would have a new income source, and the oil companies would use the octane boost that ethanol provided to reduce the octane of the baseline fuel, saving them money. In addition, state and local air officials supported the use of ethanol because it was less polluting than gasoline. For the car manufacturers, ethanol translated into a 1.5 mpg CAFE credit toward their overall mileage requirements. Automakers soon began making flex fuel vehicles, cars or trucks that can run on fuel with 85 percent ethanol content. As of 2013, there were around eleven million flex fuel cars and light trucks on the road. Unfortunately, these vehicles rarely run on anything higher than 10 percent ethanol fuel. Unlike in Brazil, higher percentages of ethanol are simply not available to most American drivers, negating the potential greenhouse gas reductions of the flex fuel vehicles.

Two years after the 2005 energy law, Congress passed the Energy Independence and Security Act, significantly expanding the volume mandate for ethanol. Instead of 7.5 billion gallons a year, the new law called for up to 15 billion gallons of corn ethanol in gasoline by 2015, twice as much as the 2005 Energy Policy Act, as well as 21 billion gallons of advanced biofuels by 2022. Once again, the law promoted renewable biofuel, but there was a problem with this legislation. I tend to look at the Energy Independence and Security Act's goals through the lens of the very different approach the EPA would have taken.

Ever since Senator Edmund Muskie's Clean Air Act, EPA regulations have required certain benchmarks with certain health benefits within a certain timeframe. Applying that set of parameters in this case, the regulations wouldn't have specified what kind of fuel should be used or exactly how much of it; they would let the oil companies sort out the cheapest, most effective way to meet the standards on a schedule. That way, EPA regulations would create new competition and markets. By comparison, the 2007 EISA law specifically required the purchase of another 15 billion gallons of corn ethanol that would be mixed into gasoline, not necessarily the best way to reduce greenhouse gas emissions.

However different our approaches, Congress did make two worthy requirements in the law. One was mandating the production 16 billion gallons of advanced cellulosic ethanol, produced from feedstocks like corn stover, switch grass, wood chips, and other non-food feedstocks, with 60 percent lower greenhouse gas emissions than gasoline. The other was that ethanol produced by plants built after 2007 have 20 percent lower greenhouse-gas emissions than gasoline.

The law also required the EPA to do a life-cycle analysis on the total greenhouse gas emissions of biofuels. Because its input material is grown in a field, gauging whether ethanol's total environmental impact is better than gasoline's depends on the associated planting, harvesting, production, delivery, and any changes in land use. We had to determine if corn ethanol blended into gasoline offered at least a modest 20 percent greenhouse gas emission improvement over gasoline. As it turns out, American corn ethanol is not nearly as effective at reducing greenhouse gas emissions as Brazilian sugar cane. I knew that, if our science showed the corn ethanol blend didn't meet that 20 percent greenhouse gas reduction goal, the powerful corn lobby and their congressional supporters would be very dissatisfied. Implementing the rule could create a politically charged situation, but my office couldn't be concerned with the politics. Doing the science properly required plenty of effort itself.

For one, tracing and quantifying the greenhouse gas emissions in the life cycle of corn ethanol required a lot of information we didn't have. We needed to include data on the emissions from the machinery that planted and harvested the corn, the trucks or trains that transported it to manufacturing facilities, the actual production process to make ethanol, and the distributing of the ethanol to blending facilities—before we even got to the impact of ethanol burning in a car. And these weren't the only factors we had to look at.

The United States is far and away the world's largest exporter of corn for food. Because the law required that a huge amount of corn be turned into ethanol, the United States would export less of the grain. Other corn-producing countries would make up for that missing commodity by growing more, either for domestic or export purposes. This is called indirect land use change. In the case of a corn-producing country like Brazil, this added corn production would result in additional land being cleared of trees and other vegetation. The EPA had never calculated the negative impact of deforestation for agricultural purposes in foreign countries. We had to gather data from a number of new sources, including NASA spacecraft that tracked historic land use changes as a result of agricultural policies.

When we finally finished our initial analyses, we found that neither corn nor soy met their greenhouse gas reduction thresholds. The ethanol and farm lobbies wouldn't like it, but we had done exhaustive research. These were the results on which the law required us to report. After we announced our findings, I was compelled to attend several mostly hostile Congressional hearings, where I was asked questions like: "Why is the EPA worrying about what's going on in Brazil?"

I was also asked if I had visited a farm. It didn't really have much to do with our analyses, but not answering questions at Congressional hearing isn't much of an option.

"My grandparents in Greece were farmers," I said. "I spent a lot of time on the family farm while I was growing up."

"Not Greece. What about a farm in the United States?"

"No." I had been to orchards in the United States, but not to any corn farms.

Soon scores of articles, letters, and emails came in asking me how I could do my job without having visited a farm. Iowa's Senator Grassley called me a "faceless bureaucrat" and publicly invited me to visit a farm. I wasn't a politician or even a political appointee, but this did not stop the attacks on me. Gina McCarthy, my boss and the assistant administrator of the Air Office, and I put on our boots and jeans and flew out to Iowa.

On the plane ride, I sat next to a very pleasant man who asked me where I worked.

"The EPA," I said.

His eyes lit up, and he smiled. "I know who you are. You're famous in Iowa. You're the woman that hasn't visited a farm."

Gina and I went to a family farm, took a ride on a harvester, and were amazed by the amount of technology used to grow and harvest crops. After my visit, no one can say that I haven't been to a farm, but the political and financial interests in promoting ethanol from corn, soy, and other feedstock remain.

★   ★   ★

To get to the necessary greenhouse gas reduction targets by 2050, a second generation of biofuels offers more exciting prospects. Some of these second generation alternatives, such as cellulosic ethanol, will reduce greenhouse gas emissions by at least 60 percent versus gasoline, a huge improvement over corn ethanol or natural gas.

The raw material of these fuels is also abundant and doesn't require cutting down food crops. Cellulosic ethanol, for example, is made from wide array of sources, including wood chips, fast growing grasses, and corn stover, the parts of the corn stalks left in the field after harvesting. There are also new technologies that biochemically convert grasses to ethanol, some that convert waste into gases that are then converted into ethanol, and still others that use algae to make fuel directly from water and sunlight. According to the Union of Concerned Scientists, there is enough biomass to produce more than 54 billion gallons of what is referred to as second-generation ethanol by 2030.[5]

So far, however, these fuels have yet to live up to their potential. The 2007 energy legislation called for one billion gallons of advanced ethanol by 2012. As of 2015, there are only a few million gallons of commercial capacity. While electric, hybrid, and even fuel cell technology are now available in production cars, alternative fuels lag far behind.

Nonetheless, when I was at the EPA, I witnessed significant advancements in enzyme and catalyst technologies, allowing cellulosic biofuel producers to achieve greater yields of biofuel per ton of feedstock. These advancements have lowered production costs from $4.00–$8.00 per gallon in 2007 to $2.00–$3.50 per gallon in 2012.

As cellulosic biofuel producers gain experience, it's likely that production and capital costs will continue to decline—just as they do with many emerging energy technologies. For example, the commercialization of the dry mill process for producing traditional corn ethanol began in 1983. Over the next twenty years, the processing costs had decreased by almost half, and the capital costs of a facility dropped nearly 90 percent. An even more rapid drop in cost barriers to producing cellulosic ethanol may be approaching. A 2013 Bloomberg New Energy Finance survey found that production cost of enzymes to create cellulosic ethanol has dropped 72 percent from

2008 to 2012. For the moment, though, second generation ethanol remains significantly more expensive to produce. It costs about 94 cents to produce one liter of advanced fuel versus 76 cents per liter for corn ethanol.

Aside from cost, ethanol faces another barrier in the United States, the politically charged dispute over the "blend wall." Oil companies argue that increasing the level of ethanol above 10 percent in gasoline will destroy existing car engines. The EPA has determined after extensive testing that, for any cars made in 2001 and later, there will be no damage from ethanol up to at least 15 percent. As with any dispute involving the petroleum industry, the blend wall issue could subject the EPA to intense political and legal pressure. However, a new technology may make the argument disappear.

Ethanol, a kind of alcohol, isn't the only form that biofuels can take. One alternative, called drop-in fuel, has a chemistry more similar to the hydrocarbons in gasoline or diesel. The fuel gets its name because conventional ethanol cannot be moved through existing fuel pipelines because of its tendency to absorb water. However, drop-in fuels can be "dropped in" to existing refining and distribution for petroleum products. This biofuel category even has the support of some petroleum companies.

A joint venture of BP, Dupont, and Gevo has developed butanol, a drop-in fuel that can be blended up to 16 percent with gasoline without any of the blend wall issues raised by oil companies. Butanol can be made from various biomass feedstocks, including corn, corn stover and other crops.

Whether or not butanol is the final answer, the low carbon fuels of the future have to be based on non-food cellulosic feedstocks, and, if produced as drop-in fuel, they can use existing refinery and pipeline infrastructure. Being realistic, I do not expect that we will meet the 21 billion gallon Congressional mandate for advanced biofuels by 2022. But with the right policy, tax incentives, and investments, the United States can lead the globe in low carbon fuel production by 2030. Accelerating our production is essential to reining in global warming. By 2050, the National Academy of Sciences reports that we will need 40 billion gallons of cellulosic biofuels complementing plug-in hybrids, fuel cell vehicles, and electric vehicles to meet the 80 percent greenhouse gas reduction goal by 2050.

★   ★   ★

In 2002 a team from Society of Automotive Engineers and the Department of Energy approached my office with—what seemed at the time—the ludicrous idea of EPA participating in a "green racing program." Because race cars are all about maximum horsepower, fuel economy had never even been an afterthought in their design. I was not sure how this could work.

I listened, though, as the teams plowed ahead with the concept of a "race within a race." Participation in the high profile racing series would challenge some of the world's best automotive engineers to reduce greenhouse gases by using clean fuels and advanced technologies.

Eventually, they won me over. I decided the race would be an effective way to not only advance the state of the art but also create greater awareness of alternative fuels. Before I knew it, I was shouting, "Gentlemen, start your engines!" and waving a little green flag at the Petit LeMans Green Races. I was even proselytizing for the event at a speech I gave to the racing audience in Georgia during the Petit LeMans race. "If you can show that clean, efficient technologies can serve extreme conditions of the American Le Mans Series without hurting performance or durability," I said, "then certainly I see no reason why the same technologies cannot work for the morning commute."

Don Panoz, best known as the inventor of the nicotine patch, owned the American LeMans Series. He and CEO Scott Atherton became ardent champions of the event. The initial target was to displace 44 percent of the petroleum used in 2005 series by 2011. The Society of Automotive Engineers developed a standard for the program called the J-2880. The Department of Energy team developed a multi-variable scoring formula that measured the cleanliness of the race car in oil use and greenhouse gas emissions, efficiency of energy use, and the speed achieved.

The program eventually became an excellent testing ground for some advanced cellulosic fuels. The race also provided an important forum for discussions with the senior technical leaders of the various automakers as they watched their experimental technologies being put to the test.

The Corvette team, which has probably the most wins in the series, runs all their races now on cellulosic E85 fuel made by Ineos Bio, a Florida-based company. By converting bio-waste from Florida's salad bowl region, Ineos Bio produces more than eight million gallons of high performance fuel. As Mark Kent, director of racing for Chevrolet said, "We have raced with E85 for many years. The reason we did was that it not only allowed us to learn things in a rapid time frame, it was also a great platform to demonstrate to consumers the viability of E85."[6]

But the battle to reduce greenhouse gas emissions by 2050 expands well beyond the breakthroughs in clean technology in the coming decades. The different attitudes toward mobility held by people in their late teens through early thirties are likely to result in greater use of technologies that connect not only people but also vehicles, and in turn enable new approaches to mobility. Fortunately, some of these technologies, when combined with the new social attitudes, stand to further advance our quest to reduce greenhouse gas emissions in ways we had not imagined.

# 18

# A WORLD OF CONNECTIONS

Connected living, a series of Internet-enabled social trends, has already changed the way people communicate and interact with each other. Instead of driving to work, more people are telecommuting. Instead of going out with friends, others are Instagramming photos of food and beer to their followers. Because online connectivity is an increasingly important form of personal mobility, younger people in the United States and Europe are less likely to have a driver's license or want to own a car. Combined with car sharing programs, connected living has the potential to make a huge difference in how people get around in the coming decades. But because these social trends are the result of a certain level of readily available technology, their impact outside of the wealthier OECD nations is likely to take a little while.

However, there is a second frontier in the new connected world, the Internet of Things. In 1999, a British technology researcher named Kevin Ashton proposed the term, which he later described as a system in which computers were no longer dependent on humans for data but could get information from other things. "[W]e would be able to track and count everything," explained Ashton, "and greatly reduce waste, loss and cost. We would know when things needed replacing, repairing or recalling, and whether they were fresh or past their best.

The Internet of Things has the potential to change the world, just as the Internet did. Maybe even more so."[1]

As intense urbanization continues, megacities will rely on the Internet of Things to make their urban networks more manageable and more livable. From energy efficiency to automatically managing traditional and alternative energy supplies, to traffic lights talking with cars to maximize flow and reduce congestion, connected things will change our environment in ways we cannot begin to imagine. Among the dramatic impacts of the Internet of Things will be making urban transportation smarter to reduce its carbon footprint.

★   ★   ★

In the late 1990s, BMW used to advertise that their new cars had more electronics and computing power on board than the Apollo 13 module that landed on the moon. Today, that could be said about the average, mass-market car. According to Paul Mascarenas, Ford's former chief technology officer, the 2013 Ford Fusion has more than 145 actuators, 4,716 signals, and 74 sensors, producing more than 25 giga-bytes of data hourly and more than 70 onboard computers analyze the data in real time.[2] Electronics have become such a big part of the total package that some industry estimates put the cost of these technologies at 40 percent of the vehicle's cost today. This percentage represents a 20 percent uptick in just ten years.

As the automobile transforms from a mechanical device into an electronic device with mechanical content, the nature of vehicle's interaction with their drivers will also inevitably change. Take, for example, the remarkable change in the relationship between people and telephones over just the past few decades.

For most of the twentieth century, the phone was a tool that peo-ple used for a specific purpose, calling other people. When no one was using the phone, it was just an object sitting on a table like a vase of flowers. Eventually, some technology expanded its features. Answering machines, for example, recorded calls with no user present, but people still had to push a button to get their messages. Even early cell phones were essentially wireless versions of the combined phone and answering machine, entirely reliant on a human user. Then the iPhone was introduced in 2007, driving the popularity of a new category of telephones. Suddenly, communication devices operated like active nodes connecting us to a much broader network.

Smartphones are still used for calling but also emailing, texting, instant messaging, mobile gaming, updating Facebook, posting on Instagram, tweeting, and accessing the nearly endless reaches of the Internet. When smartphones are outside communications networks, they actively search for signals and send and receive messages and update the weather as soon as they reconnect. They grab information from other nearby devices. They remind users of fitness goals and when to drink water or go to meetings. They record heart rates and turn a sofa into an office space. Because the cycle of innovation is so much faster in electronics than in the automobile industry, the development of smartphones provides a peek into the as yet unrealized future of cars.

Like the original phones, vehicles were a tool that drivers used for one purpose, to get from one place to another. From the 1970s, electronic features like cruise control, electronic fuel injection, and airbag systems were added to the machines. But, as useful as they were, they didn't connect the driver to additional networks. Additionally, this increasing number of electronic features continued to rely on an outdated electronic architecture instead of an integration of vehicles' electronic systems. Automobiles stubbornly maintained their system of four or five individual "buses" for different functions. The powertrain bus, for example, connects all the engine and transmission computers to ensure smooth and powerful acceleration while maximizing fuel economy and minimizing emissions. The body bus deals with the electronics that light the car and control the suspension. The safety bus links the dozens of airbag sensors, rollover sensors, and braking controls. Over decades, the increased features caused the wiring for these systems to grow heavier, but the different buses were still not linked. All the gateways were controlled, and data transfers between buses were purposefully limited.

In theory, this lack of connectivity would prevent faults, like pressing the brakes and having the horn go off unintentionally. But other transportation vehicles—including airplanes—have updated their electronic architecture without any safety issues. It is, in fact, this interconnectedness makes the pilot assist functions possible and reliable. Without connecting the lights, landing gear, acceleration, and braking controls, there would be no way that airplanes could have an in-flight autopilot feature or a feature to land and take off in inclement weather with the pilot mostly in observation mode.

The lack of internal linkages in automobiles resulted in vehicles that couldn't effectively connect the drivers with the networks of the outside world. But over the next decade, these limitations will be exposed as our expectations of cars radically change. Instead of treating them as a smarter tool, like an early cell phone, drivers will expect their vehicles to "connect" with the rest of their daily lives. Instead of oversized, luxurious go-carts, cars and trucks will be a "personal operating system" enabling drivers to experience transportation in much the same way they relate to smartphones, computers, and other networks on a daily basis. Automakers already realize the inevitability of these changes. "Now is the time for us all to be looking at vehicles on the road the same way we look at smartphones, laptops and tablets," said Bill Ford Jr., executive chairman of Ford, "as pieces of a much bigger, richer network." *Forbes*'s Detroit bureau chief Joann Muller describes the battle over the future car succinctly: "Detroit versus Silicon Valley. The war is on."[3]

As automakers scramble with this transformation, some have begun introducing their new features at the premier electronics convention, the Consumer Electronics Show in Las Vegas, instead of an auto industry event in Detroit or Frankfurt. At the 2014 Consumer Electronics Show, Audi chairman Rupert Stadler admitted that car companies were currently behind the curve: "The speed of innovation in the electronics industry is much faster than that of the auto industry . . . so we're working hard to more closely match up the production cycles and to enable easier integration of new electronics and software into existing products."

Simultaneously, electronics companies are fighting for leadership of "in-car computing." Apple's CarPlay replicates an iPhone user interface on the screen of the car. Google introduced a similar product called Auto Link for its Android customers. Cisco, the Silicon Valley–based manufacturer of networking equipment, envisions the car as "just another node on the Internet." The company predicts autos will eventually have an IP address just like on smartphones or computers. Cisco also suggests that vehicles' various internal systems, or buses, should communicate internally via Ethernet rather than through their much slower and outdated electrical architectures. Other electronics companies suggest replacing cars' fifty-pound communication wire harness with Bluetooth, an innovation that would also make the car lighter.

While connectivity inside the vehicle is still being addressed, the larger issue of connectivity with the external world will also be advancing. As it begins communicating with traffic control systems, congestion updates, and other vehicles, the car will join the Internet of Things. When I talked with Andreas Mai, Cisco's director of smart connected cars, he predicted a rapid increase in the number of vehicles connected to the Internet of Things.[4] Today just 1 percent of cars communicate with the outside world. By the early 2020s, Cisco expects that nearly a quarter of all cars will be connected, processing around 400 million gigabytes of external data per month.

Individually, all this increased connectivity and processing power will create cool and useful new features in vehicles. But the improvement of automobiles' internal and external connectivity is setting the stage for perhaps the most important innovation in the future of transportation: the driverless car.

★　★　★

In *Imperial Earth*, a 1976 science fiction novel by Arthur C. Clarke, a traveler from another planet comes to earth. While riding across a city in a driverless car, he notes the dense traffic surrounding him.

"I hope all those other cars are on automatic," he says anxiously to his host. "Of course. It's been a criminal offense for—oh—at least a hundred years to drive manually on a public highway. Though we still have occasional psychopaths who kill themselves and other people."

A year after Clarke's fictional version, the first truly autonomous vehicle was unveiled by Japan's Tsukuba University. Then things went quiet for a long time, with no major new breakthroughs. Because driverless technology is revolutionary, its mention also attracts skepticism and technophobia. Nonetheless, multiple companies—including major car companies and Google—are working to develop their own autonomous vehicles. Some of this technology is also already being introduced, but in graduated phases. Small features like cameras that help drivers stay in their lanes, adaptive cruise control that manages distance to vehicles in front and is integrated with braking, and pedestrian detection sensors to avoid accidents in urban centers have gradually appeared in new cars.

To track the growth of driverless technology, the National Highway Safety Transportation Administration has developed five levels of automation. Level 0, for example, denotes no automation, or vehicles in which the driver is responsible for all primary vehicle controls and monitoring the roadway.

At the first taste of automation, Level 1, drivers are in charge but can choose to cede limited authority over one function-specific control. The most common example of this level, cruise control, was introduced by all major American automakers as a fuel-saving option in 1974. The Level 1 label could also apply to automatic breaking or lane centering.

The Level 2 rating simply allows two Level 1 functions to be automated in unison. For example, adaptive cruise control could be selected in combination with lane centering. In some cases, a driver is disengaged from physically operating the vehicle by having his or her hands off the steering wheel and foot off the gas pedal. GM has a feature called "super cruise" that uses cameras and radar to steer the car from lane to lane, keep it a safe distance from other vehicles, and bringing the car to a full stop when necessary. The 2015 Mercedes S class, introduced as the "first generation of an autonomous vehicle," can change lanes by itself.

Level 3 signals proto–driverless automation. Under certain conditions, the car is completely autonomous, although the driver needs to be available to transition back for occasional control. A number of automakers are or will soon be offering such technologies in production vehicles. Nissan, for example, plans to deploy a trafficjam pilot by the end of 2016 and an automatic lane-changing feature in 2018. These two features will be followed in 2020 by a smartassist system to negotiate city intersections without driver intervention.

Audi also has a piloted driving system that can safely drive a car at less than 37 mph in heavy, low-speed highway traffic (Figure 15). The system includes forward-looking radars, a laser-ranger to monitor traffic ahead, a windshield-mounted camera to follow the lane lines, and a rear radar to detect cars advancing on either side, which

**Figure 15. Audi Autonomous Driving Dashboard**

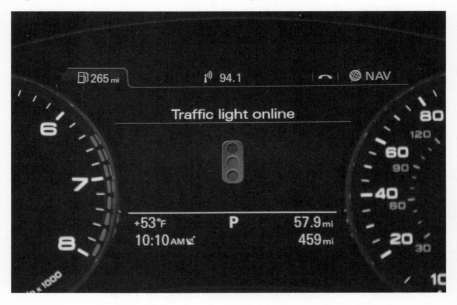

(Source: Audi)

keeps the vehicle in the lane at a set distance from the one ahead. The company is also planning to practice communication with traffic light control systems in several test-site cities in Germany to reduce wait times at lights.

From the Level 3 technology, which will be available at dealerships by 2017–2018, it is a surprisingly short step to a fully driverless car. In these Level 4 vehicles, the driver will provide destination input but is not expected to be available for control at any point. In fact, the cars could be unoccupied. Most of this prototype technology is still in the testing phase. At the 2013 Consumer Electronics Show, for example, Audi demonstrated a self-parking car. After exiting the vehicle, the passenger uses a smartphone to tell the car to go park itself at a nearby garage.

Mercedes has also demonstrated a Mercedes S class that drove itself nearly eighty miles through Germany without driver input, retracing the same route taken by Bertha Benz 125 years earlier in her husband's car. A 40-ton Mercedes truck drove fifty miles per hour through an unopened highway in Germany without driver input. The truck, like

**Figure 16. The New Google Vehicle**

*(Source: Google)*

the car, demonstrated the ability to stay in a lane while accelerating and slowing down for traffic.

Despite the impressive show of technology by the German and Japanese automobile manufacturers, it is the work of a Silicon Valley tech company that has captured the world's imagination. As of mid-2014, Google's fully autonomous, driverless car fleet of modified Priuses and a Lexus 450h had logged some 700 million miles without an accident on traditional roads with normal traffic. Then, integrating its extensive computing and mapping capabilities, Google took a surprising step: building its own car (Figure 16).

The Google vehicle's speed is limited to 25 mph, thus avoiding the typical certification requirements from EPA or the Department of Transportation. It seats two passengers, has no steering wheel, and is electrically powered. According to Chris Umson, director of Google's Autonomous Vehicle project, the car has independent safety features like an emergency stop button as well as a soft front end and flexible windshields that will reduce any impact if a car does hit a pedestrian. The car does not change lanes and can't be operated in dense urban traffic yet. In short, Google's vehicle is more of an electronic device than a traditional car, but its computing capabilities are enormous.

The Google vehicle has a high-definition laser system on the roof that captures 1.3 million bits of data per second, along with video feeds and radar pulses. To integrate sensors that give the car 360-degree vision, the surface of the car is curved. Onboard computers can "see" the objects ahead and catalog them as another vehicle, a person, even a bicyclist making a hand gesture to make a turn. Integrated with Google's 3D mapping data, the vehicle is continuously adjusting to its environment.

Google cofounder Sergey Brin, who views self-driving cars as an on-demand service, has set the goal of commercializing the technology by 2017. Coincidentally, that is also when his children reportedly reach driving age. If successful, autonomous vehicles could significantly dent future auto sales, especially for the current market leaders. The financial services firm KPMG found that American consumers would trust brands like Google and Apple more for self-driving cars than they would automakers.

Whoever ends up manufacturing driverless cars in the next decade, the widespread appearance of the technology is so certain that the National Highway Safety Administration formally defined autonomous vehicles as "those in which at least some aspects of a safety-critical control function (e.g., steering, throttle, or braking) occur without direct driver input." As envisioned in the illustration in Figure 17 by the DOT, the automation is enabled not only through technology on board but through interaction with other vehicles as well as the infrastructure around the vehicle.

Some states are already welcoming autonomous vehicles on to their roads. On March 1, 2012, Nevada began allowing operation of driverless cars without specialized permitting. Later that year, Florida and California followed, each with their own restrictions and permitting

**Figure 17. DOT Illustration of the Connectivity for Autonomous Vehicles**

*(Source: DOT)*

processes. In 2013, Michigan not only allowed driverless car operation but set up a thirty-two-acre mock city on the campus of University of Michigan in Ann Arbor to test driverless cars. On the course, there are buildings, road signs, construction sites, and even "mock pedestrians" that can rush into traffic.

But it's not only in cars that we see autonomous features. In my discussion with Tana Utley, a vice president of Caterpillar, she indicated that a number of her company's products include semi-autonomous features that take certain operations out of the hands of machine operators, using electronic controls to improve both efficiency and performance.[5] The electronic controls are much more efficient at, say, operating a blade than a human operator. In the mining industry, Caterpillar offers a 400-ton mining truck that operates "autonomously," allowing the operator to monitor performance from a control center. This not only helps improve productivity—mining trucks do not require as many breaks as a human operator—but addresses safety and a shortage of skilled workers interested in working in hazardous, remote locations operating heavy equipment. The truck also uses less fuel and thus creates fewer carbon dioxide emissions.

In Europe, Germany, Holland, and Spain also have laws allowing driverless cars to be tested on selected roads. The Dutch are particularly focused on commercial trucks, given the importance of Europe's largest port, Rotterdam, in the continent's shipping trade. Their vision is to have driverless trucks delivering goods from Rotterdam to select destinations by 2020. I believe that many cities and states will soon go beyond simply allowing driverless cars and trucks on their roads to requiring that every vehicle driven within city limits have some level of autonomous features.

★   ★   ★

In the future, megacities will continue to find ways to clean up their air, reduce congestion, and limit greenhouse gas emissions. The driverless car accomplishes all of these goals at a much lower cost than alternative transportation measures like expanding subway lines.

Autonomous vehicles may, for example, help to reduce the vehicle fleet—significantly. In a traditional commuter situation, a car could drive one parent to work and be available for use by the family until it was time to pick up the parent. Likewise, combining car sharing with vehicles that delivered themselves autonomously would further reduce dependence on owning a car.

Autonomous vehicles would also save lives. Every year, more than 30,000 Americans die from traffic accidents. Ron Medford, the former executive with NHTSA and now the director of safety for Google's self-driving car project, told me that he expects self-driving cars to be "one of the most important life-saving technologies after safety belts and electronic stability control."[6]

There are numerous other inefficiencies that driverless cars would reduce. For example, in 2006, drivers stopped at a traffic signal in the SoHo district of Manhattan were interviewed for a survey. A full 28 percent said they were cruising for curb parking. A similar study in Brooklyn found that number to be 45 percent. Driverless cars can simply drop off passengers without worrying about parking.

From a purely financial standpoint, autonomous vehicles' potential savings for city governments and residents—as well as the planet as a whole—are immense. In 2011, Cisco estimated that the "real cost" of vehicle ownership in the United States was in fact $3 trillion, two times the "actual cost" of $1.5 trillion, with the rest being losses from congestion, crashes, and extra fuel. Supporting Cisco's argument, Morgan Stanley calculates the potential value—savings and benefits to US society—at $1.3 trillion, very close to the Cisco number. Among these savings was $488 billion saved annually from reducing the number of accidents, with another $169 billion in fuel savings from better traffic flows and reduced congestion.

The fuel saved also has health dividends. Using EPA's numbers,[7] the predicted $169 billion would translate into roughly 375 million tons of carbon dioxide saved annually. This could be a significant future "multiplier" for the needed carbon dioxide reductions beyond 2025.

Autonomous cars will also have a big impact on plug-in electric vehicles. As Robbie Diamond, president of the Electrification Coalition told me, "I believe that self-driving car is going to very quickly change the world. Batteries in an electric car no longer will need to go 200 miles. . . . You just get sent a car with the right battery, and it will be inexpensive, and the car companies will essentially own the cars. If you have an Uber application, where you call your car and tell them where you're going, they're going to send you a car with the right battery to drop you off that place and then drive over to an inductive charger that will charge within minutes, and then it will go to the next place. . . . It will totally change the way we use cars, see cars, own cars, fuel cars."[8]

The possibilities are real enough to concern the auto and insurance companies. PWC's Autofacts, auto industry forecaster, released a provocative paper suggesting that self-driving cars may eventually reduce the number of cars on America's roads by 99 percent, from 245 million to as little as 2.4 million.[9] The insurance companies would see the same proportion of their premiums disappear. This transformation would also eliminate most of the fuel needed by today's transportation sector, an enormous step toward greenhouse gas reduction targets.

★   ★   ★

Because of its transformative nature, autonomous vehicles open up all sorts of new questions. David Strickland, the former NHTSA administrator, pointed out the issue of liability: "Who is going to be responsible? Because we have this model where the driver's responsible, but if it's something that the vehicle is doing then the manufacturers are responsible for that." Another concern with a vehicle that has so many linkages to the outside world is cyber security. In a recent episode of the HBO comedy *Silicon Valley*, an autonomous vehicle picks up a character, but the car soon changes directions and drives itself—and its distressed passenger—to be loaded into a shipping container bound for an island that straddles the International

Date Line. This technophobic comedy nightmare is, in fact, a real possibility.

In 2010 several Rutgers students demonstrated how a car could be hacked through its wireless tire pressure monitoring system, a safety requirement since 2008. Tesla Motors also admitted that one of its cars had been hacked in China. The Internet security company Qihoo 360 Technology found ways to remotely control the Tesla car's locks, horn, headlights, and skylight while the car is in motion.

Though the range of concerns is broad, some are justified. The notion of someone else "hacking" in and taking control of a car for the fun of it, or with a deliberate purpose, has been the subject of debate in numerous conferences and tech magazines, like *Infotech*. As most of the traffic will flow over the Internet, the vehicle can be "overloaded" the same way websites are. Hacked driverless vehicles are also a dream for thieves, who can send the car to remote location to be picked up later. Such vehicles would also be enticing for terrorist to deliver bombs or for other criminal activities.

Cisco is already promoting its Auto Guard Cloud security services for connected vehicles, arguing that all their traffic should be routed through cloud where existing filters that protect Internet traffic could be adapted to also protect the communication traffic with the vehicle. Companies will develop various solutions to safeguard connected vehicle technologies, but as with Internet viruses, there will likely be hiccups along the way.

There are also questions about the limits of the environmental impact of autonomous vehicles. Researchers at Toyota argue that driverless cars could actually increase vehicle miles traveled. As commutes become less burdensome, they say, people will be willing to travel longer distances. Extending commutes would definitely offset some or all of the fuel and emissions savings of autonomous vehicles.

Finally, there is the fundamental question of consumer acceptance. Drivers may trust Google's driverless car more than Ford's, but that doesn't mean they are going to buy either yet. A Harris Poll found that just 12 percent of Americans wouldn't be concerned about turning

over driving duties to their car.[10] More than half said they were worried about hackers getting into the car's computer, and 79 percent
questioned whether the technology might fail at some point.

Though autonomous driving opens many questions, the technology also provides answers for transportation problems that wouldn't
have been on urban planners' radar a decade or two ago.

# LESSONS AND PATHWAYS TOWARD 2050

Last year, a campfire on a drought-stricken slice of the Pacific Coast was fanned by seventy-mile-per-hour winds into a wildfire that burned down hundreds of acres of pine forest. The event was not particularly remarkable—out-of-control fires are a natural feature of the western landscape every summer—except for two details: The fire started in late January, and it raged through the coastal ranges of Oregon, one of the wettest winter regions in the continental United States. That January, the site of the fire had only a third of its average number of rainy days and 80 percent less precipitation than normal.

The weather on the rest of the Pacific Coast was equally bizarre. To the south, Los Angeles and San Francisco had their warmest and driest Januaries ever. To the north, heat closed ski resorts in Alaska, and the town of Port Alsworth matched a statewide January high of 61 Fahrenheit (16 degrees Celsius). Ten days later, the temperature in Howard Pass dropped to an all-time Alaska wind–chill-adjusted low of –96 Fahrenheit (–71 degrees Celsius). Frigid weather was more prevalent in the rest of the country. Detroit dug out from under its heaviest snowfall since 1880. Compacted snow, ice, and sleet interrupted the normally mild Southeastern winter, causing Atlanta commuters to abandon their cars and walk miles to avoid up to twenty-hour delays in traffic.

Every year has its own extreme weather events. What's more, climate scientists are careful to avoid linking any one specific event to climate change. But the science is clear on the bigger picture: we are already living in the era of global warming. We have been for decades. Extreme weather events from intense heat and droughts are going to become more common and change the way we live. Climate change is caused by the buildup of human-produced greenhouse gases in the atmosphere. But humans have mostly avoided facing up to these facts, instead choosing to kick the problem down the road.

That same January, England endured its second wettest December and January since 1766. London responded by raising the Thames River flood barrier a record eighteen times. The same month, tennis players at the Australian Open vomited, fainted, hallucinated, and had the soles of their shoes melt on the court in temperatures of up to

**Figure 18. The Elephant in the Room: GHG Reductions Required by 2030 and 2050**

**Total Global GHG Emissions**

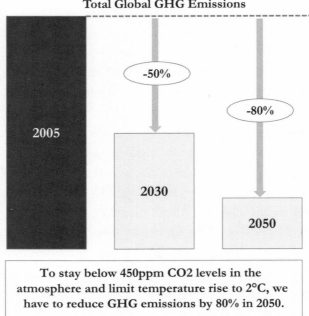

To stay below 450ppm CO2 levels in the atmosphere and limit temperature rise to 2°C, we have to reduce GHG emissions by 80% in 2050.

*(Source: IPCC Analyses)*

108 degrees. The tournament went on, with players lying between ice towels during set breaks.[1]

In the coming years we will have no choice but to engage in both short- and long-term mitigation of the effects of climate change. But if we don't also accept the bigger task of preventing runaway global warming, our flood barriers will eventually give way and iced towels will not prevent hundreds of thousands of deaths in heat waves.

There is, in turn, only one certain way to prevent catastrophic global warming: massively reduce the carbon dioxide emissions from all sectors of the global economy. Because the United States is responsible for around 25 percent of these emissions, we have a responsibility to act and to lead the world by our example. I'm arguing for a large reduction in greenhouse gas emissions from transportation: 50 percent by 2030 and 80 percent by 2050 from 2005 levels (Figure 18). After spending eighteen years working with every major player in the transportation sector, I believe that the cars of the future will help us meet these goals.

★   ★   ★

I can't predict precisely what car my grandchildren will be driving in 2050. But elements of it may be more like a Toyota concept car called the Fun Vii than anything on the road today. The vehicle lives up to its name. It features a "digital skin" that allows drivers to select colors and upload photos that cover the interior and exterior of the vehicle (Figure 19). A 3-D hologram navigational attendant answers operational questions. And the car's whole interior can essentially be turned into a giant video game. But these features are just fun—very fun—eye candy.

The car runs on a pure electric engine that can be recharged wirelessly. Its name stands for Fun Interactive Internet, so it's not surprising that the vehicle is electronically networked with other vehicles on the road, an important capacity in crowded cities. The car also features a steering wheel that tucks away when in driverless mode

**Figure 19. Toyota Fun Vii Concept Car**

*(Source: Toyota)*

and sleek aerodynamics, both of which create fundamentally more efficient operation.

All these technologies still sound futuristic but, unlike the flying cars promised in the 1950s, are realizable in fairly short time frames. What's more, Toyota does not own any of these ideas—they are being pursued by multiple companies. In other words, the Fun-Vii is not

**Figure 20. BMW i8**

*(Source: Shutterstock)*

just a pie-in-the-sky concept car but a creative convergence of technologies already underway. Truly futuristic cars are the result of multiple dynamics working together.

One such dynamic is a marketplace demand for cars that are not just small, efficient, and purely functional. Tesla, a company on the vanguard of auto innovation, also seeks to satisfy high-end auto enthusiasts with their battery electric vehicles.

The traditional luxury automakers will inevitably continue to innovate. The BMW i8 just hitting the markets is a carbon and aluminum plug-in hybrid electric that generates 350 horsepower from a combination 1.5 liter gas engine and electric motor, goes from 0 to 60 mph in under 4.5 seconds, and has an EPA rating of 76 mpge, which translates into less than 80 mpg in carbon dioxide emissions (Figure 20). It costs about 20 percent more than a 260-mile-range Tesla S but sold out in the UK by mid 2014. The current wait time was nearly a year in late 2014.

A few years ago at the Los Angeles Auto Show, Mercedes showed its Biome car, a vehicle that completely reimagines not just the materials but the process with which cars are built (Figure 21). The Biome concept is built around BioFibre panels harvested from organically

**Figure 21. Mercedes Biome Project Concept**

*(Source: Mercedes)*

grown, genetically modified trees. BioFibre would be lighter than metal or plastic, yet stronger than steel. A Mercedes car made of this material could weigh in at 875.5 pounds, about the same as the Harley-Davidson motorcycle.

As impressive as that weight reduction and the resulting improvement in greenhouse gas emissions are, the manufacturing process is its most amazing feature. According the company, the interior of the Biome literally grows from the DNA in the iconic Mercedes star on the front of the vehicle. The exterior grows from the star on the rear. The wheels would be grown from four separate seeds. Because of the radical departure from traditional assembly line production, Mercedes would be able to accommodate specific customer design requirements by merely changing the genetic code before it is combined with the seed capsule. Not surprisingly, the vehicle doesn't run on any traditional fuels but a product called "BioNectar4534," which is stored in the car's BioFibre structure. The only by-product of the fuel is oxygen. The entire vehicle would also be completely biodegradable.

None of the materials for the Biome project are commercially available, but the revolution beginning in transportation is already making distant thoughts realities. Ten years ago, how many people would have imagined a battery electric vehicle that was faster, safer, and more fuel efficient than all but a few other cars on the road? Five years ago how many people outside of Google were seriously discussing driverless cars? In a transportation revolution, the cars of the future are made of dreams like these. But there are still several very real obstacles to overcome before we get there.

★   ★   ★

First of all, the 80 percent greenhouse gas reduction goal requires some boundaries for consideration in imagining tomorrow's vehicles. The ideal future car—or future fleet—would be a mixture of plug-in electric vehicles powered by clean and renewable grid electricity, fuel cell electric vehicles powered by hydrogen made from renewable sources, and a very limited number of internal combustion engines powered by low carbon fuels. Remember, this is not the ideal future car in showrooms in 2050 but the range of vehicles driving on the road by mid-century. Given the typical fifteen-year average for a vehicle's useful life, cars and trucks with 180-mpge capability need to be offered by 2035 so that the fleet could turn over by 2050.

Any rapidly transforming field is necessarily going to be very complex and difficult to predict. There are so many new, innovative solutions being developed today that we, as regulators, never imagined. The potentially huge impact of driverless cars or shared vehicles is not in the "technology roadmap" of any current regulation for transportation greenhouse gas reduction. If driverless cars and trucks fulfill their promise, they can work wonders in the increasingly urbanized societies of the future. Combined with vehicle sharing, fleet sizes can be reduced, thus accelerating the reduction in greenhouse gas emissions from cars and trucks.

And the cars of the future will likely be more integrated with other power sources, further helping displace carbon emissions. Using the battery packs in electric vehicles for the storage of grid electricity is not far from reality, yet is another innovation that is missing from the regulators' "technology roadmap." Vehicle-to-Grid, or V2G, is the science of storing wind energy produced at night or solar energy produced during the day in the batteries of an electric vehicle to not only power the vehicle, but reduce peak-load power generation. This technology is being tested in numerous sites around the world. Fuel cell vehicles can also serve a similar purpose as the fuel cell in your car can keep running and creating electricity for the owners' house, or as part of the microgrid set up for the community of which they are part.

What's more, I am certain that a new market-disrupting innovator like Google or Tesla will capture our imagination and reconfigure the transportation picture ten or fifteen years down the road. And it could quite likely be a tech or electronics company rather than a traditional automobile manufacturer.

I have spent my professional life as a passionate believer in the power of innovation, and this is as formidable a challenge as we can imagine. At the EPA, we successfully kick-started the process of reducing greenhouse gases with the 2012-2016 and 2017-2025 rules that cut carbon by 50 percent and doubled the nation's light duty vehicle fuel efficiency to 54.5 mpg by 2025. As we approach 2025, the improvement curve averages 5 percent reduction of greenhouse gases per year, undeniably a significant amount.

Unfortunately, as the chart in Figure 22 shows, those improvements still only take us part of the way for the car and light truck fleet in the United States. The chart compares two scenarios. The NAS Report "Committee Reference" case scenario assumes that the 2025 GHG emission rule is fully implemented, but that there are no rules that require reductions beyond 2025. That puts us on a solid trajectory

**Figure 22. Impact of Continuing a 5 Percent/Yr Reduction in GHGs for Light Duty Vehicles from 2026 to 2050**

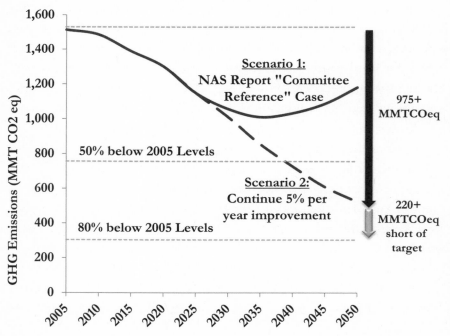

*(Source: John Koupal of ERG, Inc.)*

through 2025, but after that the reduction trend reverses as factors such as vehicle miles traveled continue to rise.

The second scenario suggests that we do not stop in 2025, but instead the United States implements additional rules that continue the 5 percent annual greenhouse gas emissions reductions, uninterrupted, from 2026 through 2050. As impressive as the continued reductions would be, they would unfortunately take us only to a 65 percent reduction level, leaving us short of the 80 percent target. We would be short roughly 220+ million metric tons of carbon dioxide equivalent (MMTCDE).

Given this analysis, it is critical that we not lose our hard-won momentum. At the very least, we must continue the 5 percent annual improvement in greenhouse gas reductions for cars and light duty trucks beyond 2025. We need aggressive targets to drive the market-based innovation necessary to meet the 2050 goals.

Seen in this light, stepping backwards even from the future car's current improvement curve would be devastating. In 2018, the current standards will be up for review. Some industry figures are already talking about the need to push back. On August 5, 2014, at a Traverse City, Michigan, car industry conference, a National Automotive Dealers Association economist suggested that the greenhouse gas regulations will stymie industry after 2018. It's "tough to think that consumers willing to pay $3,000 to $7,000 more for a car, just because someone in Washington, DC, or California says they need to."[2] The association—the same group that tried to scuttle California's Pavley greenhouse gas bill back in 2003—has refused to accept our assessment of vehicle costs in 2025. In doing so, the dealership association took a hard line, despite the support of the automakers for our 2025 standards. Our research shows that the 2025 rules will cost $1,800 on average for a new car, with a consumer benefit of up to $7,400.

The setback of falling off our current improvement curve, rather than moving forward, has significant consequences. We've already seen what happened in the mid-1980s when President Reagan relaxed the CAFE standards and fuel efficiency flatlined for two decades. The automakers may decide they can satisfy Wall Street by going back to heavy internal combustion cars rather than continuing to invest in advanced powertrains. I believe that given the global efforts to address climate change, substantial investments by car companies, and the consumer preference for fuel-efficient cars, history will not repeat itself.

Clearly, any future regulation will have to create not only incredibly clean vehicles but cars and trucks that people want to buy. I recently spoke with Tana Utley, vice president of large power systems at Caterpillar Inc., about the benefits of environmental regulations—challenging engineers to design innovations—and their disadvantages—equipment manufacturers struggling to balance rules with customer value. There are plenty of stories where both objectives were met. For the non-road emission regulations, Caterpillar not only met the environmental regulations to reduce particulate emissions and nitrogen oxides but incorporated technologies to reduce the fuel consumption and fuel cost of a medium wheel loader by 10 to 15 percent. The new technologies helped Caterpillar gain market share in the United States as well as in developing markets where efficiency and performance translate directly into improved return on investment.

Finally, the targeted reduction in emissions is one that the entire world would have to meet, not just the US transportation sector, despite our disproportionate contribution to the world's greenhouse gas emissions. When fighting global warming, everything is connected. For example, this book focuses almost entirely on efforts to reduce greenhouse gas emissions from the transportation sector, but if the electrical utilities don't create a low carbon grid to power electric vehicles, our efforts in addressing climate change will fail. Likewise, China recently implemented aggressive targets for greenhouse gas reduction from vehicle emissions. Because China is the world's largest auto market, this is an important first step. But the country is also the world's largest emitter of greenhouse gas emissions. If China is unable fulfill its recent 2020 target of reducing carbon intensity from all sectors by 40 to 45 percent from 2005 levels, global efforts to curb the impacts of climate change will be tremendously set back.

★　★　★

Meeting these challenges requires all sorts of expertise, dedication, and, most important, radical innovation in technology and how we live and work. But as varied and wide ranging as these advances are, all of them can be reinforced, encouraged, and guided by policy and government action at all levels. Cities can support car sharing, public transit, biking, and sustainable development. In fact, cities large and small and local governments will be at the forefront of the required social innovation. Former New York mayor Michael Bloomberg led the charge with his active role in the C40 coalition of cities who have taken on environmental concerns as a priority. Whether it is with the shared bicycle program in New York and many other cities or Paris's new shared vehicle program, all will have a major impact on how much we emit. In 2014, the United Nations appointed Bloomberg as a special envoy for cities and climate change, which will hopefully give him a broader platform to push his reforms.

States can provide incentives for clean energy development and incentives and infrastructure for EVs, PHEVs, and fuel cells; they can encourage novel technologies like the autonomous car. Others can push innovations, as California has done in providing exceptional leadership to lower greenhouse gas vehicle emissions and with the ZEV mandate. The federal government can fund centers of innovation and negotiate with carbon-heavy national industries and companies. Meanwhile, the EPA can continue to do what it does best: work closely with California, NHTSA, and industry in a collaborative fashion to set aggressive but achievable goals to reduce greenhouse gas emissions and then let innovative companies develop the best solutions.

This type of policy creation is where my experience and expertise sit. My eighteen years as head of the EPA's Office of Transportation and Air Quality coincided with the second half of a forty-year EPA-led revolution to clean up the nation's air. During my tenure, my team took ten major regulatory actions to safeguard the health and environment of Americans. Most of the rules targeted one or more of what are known as "conventional pollutants," such as particulate matter (soot that poisons lungs); nitrogen oxide and hydrocarbons (ozone and smog); and sulfur dioxide, carbon monoxide, and air toxics like benzene.

Mechanically, however, each rule was different. Some involved the use of specific component technologies on a vehicle, like electronics to better manage fuel to air ratios on a car or the addition of a catalytic converter or a particulate filter for trapping the small particles that are found in the black smoke of diesel cars, commercial trucks, and agricultural and construction equipment. Another one required oil companies to clean up their gasoline or diesel products, causing them to invest in new refinery technologies. But each delivered extremely cost-effective and far-reaching health benefits. And, though they were all complicated measures to put together, we were able to bring forward the learning and experience gained from previous actions. Among the most significant measures were:

### 1999 Clean Cars and Clean Fuels Program (Tier 2)

Cleaning fuels started with the removal of lead in the 1970s, a recognized health threat and impediment to the effective operation of emissions control devices like catalytic converters, long before the same was done in Europe. The 1999 Clean Cars and Clean Fuels Program continued that effort, targeting not only harmful pollutants like nitrogen oxides and hydrocarbons, but also sulfur in gasoline that damages catalytic converters, leading to increased emissions. This program prevents 3,500 premature deaths a year and tens of thousands of cases of bronchitis and respiratory illness annually. It was also the first rule in which that the EPA looked at cars and fuels as a system. Designing a regulation that effectively regulated two industries, auto and petroleum, added extra chairs at the negotiating table. But in the end, we were able to bring together the car companies, the BP oil company, the environmental groups, and the states to create a collaborative solution.

## 2000 Clean Truck Diesel Rule

Every adult can picture an 18-wheeler rolling down the highway with a sooty cloud pouring out of its smokestack. If they live in urban areas, adults can also remember stepping away from the curb as a public bus pulled away, leaving behind its own dirty cloud. But thanks to this rule, the deadly soot that causes premature death and respiratory illnesses, and can lead to lung cancer, is no longer commonplace for children and adults. As a result, 8,300 premature deaths, 300,000 asthma attacks, and more than 300,000 cases of respiratory illness in young people are prevented annually. Because this regulation resulted in America's refineries producing much cleaner diesel, it also allowed once-filthy diesel cars to apply the needed technologies to meet the federal and California emissions standards and come back on American roads.

Although the refineries were compelled to spend billions of dollars retooling their processing, they ultimately realized a net gain. Because of this regulation, diesel engines became known as "clean diesel," and the refineries that produced this fuel became profitable exporters of clean diesel to other markets, which began a broad transition to the cleaner fuels.

## 2004 Non-Road Diesel Rule (Farming, Construction, and Industrial Equipment)

After we had finished regulating buses and heavy trucks, we looked at the dirty diesel being used by tractors, bulldozers, and other farming and industrial equipment. With the on-road diesel rule as our template, we were able to create a coalition of industry, refining, state, and environmental interests to champion a commonsense solution to a

growing pollution source. This regulation will prevent 12,000 premature deaths, 15,000 heart attacks, and tens of thousand of respiratory
illnesses annually.

## 2007 Locomotive and Marine Regulations

My office at the EPA was responsible for monitoring trains, motor-
boats, and ferryboats too. The same trail of black smoke once identified with large trucks also plagued the engines that propel our trains
and boats. For the first time, we required these sources to use the
catalytic converters and cleaner fuel requirements established for cars
in the 1970s. Although these modes of transportation have to meet
somewhat less stringent requirements than cars and trucks, this regulation prevented 1,400 premature deaths and tens of thousands of
respiratory illnesses annually.

## 2009 Oceangoing Vessels

Finally, we looked at the pollution streaming behind container ships,
cruise boats, and other large vessels. Surprisingly, much of their pollution happens on land, or very close to land, dirtying the air in major
ports close to urban centers. The most amazing part about this effort is
that we had to convince fifty-two countries like Panama, Cyprus, and
others to agree to global regulations. The United States led an unprecedented effort that resulted in a global regulation to which all countries have agreed. More than 90 percent of the oceangoing vessels
that come to the United States are foreign-flagged, and it was unclear
whether, under the Clean Air Act, the United States could require
other countries to reduce emissions from their vessels as they entered
US ports. My office took a proposal to the International Maritime
Organization in the early 2000s, and it was voted and accepted by this
international organization. All member countries are now bound by
its authority. This rule prevents 13,000 premature deaths, 20,000 heart
attacks, and 200,000 asthma attacks in kids annually.

★   ★   ★

These rules have been incredibly beneficial in preventing illness and premature death among Americans. Collectively, they will have prevented 40,000 premature deaths, hundreds of thousands of respiratory illness, and hundreds of thousands of childhood respiratory cases annually at a cost of $15 billion and net public health benefit of $290 billion. This is a disproportionately huge success. The cost-to-benefit ratio alone shows the incredible success of these combined efforts. As of 2004, the Office of Management and Budget calculated that the OTAQ rules accounted for 57 percent of the public health benefits from all major government regulations put together.[3]

But the significance of these rules is not limited to their overwhelming health and welfare benefits. The experience we gained, the team we built, and the skills we honed from 1994 to 2009 served as the proving ground for our ability to deliver the success of the three groundbreaking greenhouse gas regulations for cars and commercial trucks that we began work on in 2009. For all successful environmental regulations, we have to effectively collaborate with many stakeholders. In the case of the greenhouse gas rules, we brought together California and other states, the auto industry, the Department of Transportation, environmental NGOs, unions, energy security experts, and other stakeholders to create strong new restrictions on greenhouse gas emissions.

Another measure of the rules' significance is how they defined the environmental legacies of each of the last three presidents. In July 2013, for example, President Clinton celebrated the environmental accomplishments of his administration during a ceremony renaming the EPA's Washington headquarters the William Jefferson Clinton building. But only one of the accomplishments he touted was responsible for saving thousands of lives and measurably reducing respiratory illnesses and asthma attacks among children: the clean car and clean fuels program created by my office.

During the 2004 presidential debates, Senator John Kerry pressed President George W. Bush to list his environmental accomplishments.

The one action President Bush came up with was, again, one of our office's, a rule that cleaned up the fuel and engine emissions for agricultural and construction equipment.

Finally, President Obama's signature environmental accomplishment for his first term is without a doubt the greenhouse gas rules doubling fuel economy standards by 2025. These programs are testament to the work of my extraordinary team of public servants who, collaborating with all stakeholders across the industry, states, and environmental NGOs, produced exceptionally positive health and economic benefits for Americans.

For future presidents, great environmental successes will no longer be measured by cleanups of these conventional pollutants. As important as these efforts have been, the biggest challenge ahead of us is preventing the global catastrophe of runaway climate change. Though I am no longer a regulator, I was in the middle of the first federal efforts to combat greenhouse gas–driven climate change. I hope these initial efforts lead to a global revolution to limit greenhouse gas emissions. I also know at least some of what it will take to accelerate our battle against global warming.

One of the victories of the first greenhouse gas rules was that it was an agreement between the EPA, California, NHTSA, and the automakers, thus avoiding a legal challenge. The car companies—and every other regulated industry—usually work every angle of the legal system to stall implementation of new rules. In the face of opposition by deep-pocketed industries and their teams of lawyers, regulation can be held back for years or never even implemented. In this case, we got thirteen of the automakers to agree to the rules before they were even announced, much less finalized. That is why President Obama could stand on the podium with all the CEOs to announce the rule.

The ability to walk a fine line between producing weak regulation and being sued for what stakeholders claim are overly aggressive rules was a skill we had developed prior to the greenhouse gas regulation. Earlier rules, such as our 2004 Non-Road Diesel Rule, was relatively aggressive but had remarkable success in avoiding lawsuits. At the time, our new EPA administrator, Mike Leavitt, was amazed by how our regulation managed to bring together all of the stakeholders, even the environmentalists, to support a Bush administration rule. Later, the rule became the basis of a study by the nonpartisan, nonprofit organization Council of Excellence in Government Leadership. The council highlighted some techniques it thought could help revitalize and transform how the government works.

Reflecting on the council's analysis, I've come up with my seven lessons on effective collaborative rulemaking that helped us push through the first national greenhouse gas regulations. These lessons learned are constructive guidelines but not necessarily a blueprint, for every new regulatory effort has its own players, dynamics, and legal basis for action. As we move into a world shaped by future trends like megacities, the possibility of peak oil, the confluence of global greenhouse gas regulations, and connected living, I think they will continue to be helpful in the fight against global warming.

## 1. Create long-range, performance-based stretch goals, but allow flexibility on the means.

At the start of the Clean Truck Diesel Rule, I had a long discussion with my team on where exactly to set the target for the amount of allowable sulfur in diesel fuel. Sulfur, which occurs naturally in crude oil, exceeded 300 parts per million (ppm) in commercially available diesel fuel at that time. Because the high sulfur levels fouled catalytic converters, diesel engines could not be fitted with catalysts and were spewing pollutants that were causing the premature deaths, severe respiratory illnesses including chronic and acute bronchitis, and asthma attacks. The Clean Air Act allowed us to reduce the level of sulfur to a point where catalysts could be added to diesel vehicles as they had been for gasoline vehicles since the early 1970s.

The oil industry's general consensus was that the best we could do was 50 ppm, a more than 80 percent reduction in sulfur. But based on health impact data presented by my staff, I still wasn't convinced that the number was low enough. We pushed back on the oil industry and put forward a target of 15 ppm, a proposal that won the support of the Clinton administration.

That level was far below what oil industry expected. It was a true "stretch" goal, one that could not be achieved with incremental improvements but would require pushing the technology toward the edges of the achievable. In fact, when we announced our goal in a meeting, some of the oil company negotiators thought they'd heard "fifty" instead of "fifteen." When we reaffirmed the correct target levels at meetings, some representatives did double takes. However, we made it clear that while our aggressive targets were fixed, we were willing to offer other flexibilities. We weren't going to tell them how to achieve these levels, and we would negotiate on the timeline. We gave additional time to numerous small refiners. We offered the larger operations five years to build the capability, and in addition we allowed them to sell 20 percent of highway diesel fuel at 500 ppm through 2009.

Other flexibilities included an averaging, banking, and trading. For example, a company with multiple refineries could average among its facilities or buy credits from another company to meet the company's overall regulatory requirement.

We offered the diesel engine manufacturers significant flexibilities to meet the nitrous dioxide and particulate matter standards. We didn't negotiate the level of the standard, but we gave them lead times of up to ten years and allowed them to phase in the standards by 50 percent in 2007 and by another 50 percent in 2010.

When it came to the non-road diesel, we negotiated with the oil industry to phase in the sulfur standards. In the first couple of years, the sulfur had to be at 150 ppm. Two years later, it had to meet the final goal of 15 ppm. The states and environmentalist were very concerned about this stepped approach, but, after numerous meetings with our technical staff, they realized that it was possible to get the reductions that the rules set out to achieve. In addition, we would get the oil industry support, something that was unheard of at the time.

We applied some of these same flexible strategies when negotiating the 2025 greenhouse gas regulations. For example, automakers had to meet the aggressive new standards, but each company had to worry only about its own "custom" fleet average, which reflected the typical mix of vehicles they produced. The more efficient cars and light duty trucks in their fleets could offset ones that didn't meet the standards. This meant the companies didn't have to stop producing any one specific model. Similarly, companies faced less stringent standards for trucks, heavier vehicles for which it is more challenging and expensive to reduce greenhouse gas emissions. Our biggest incentives may have been the additional credits we gave to companies investing in game-changing technologies like EVs, PHEVs, and fuel cell vehicles. We valued the investment in and manufacturing of EVs so much that we essentially offered automakers a two-for-one deal. For every EV manufactured, the automaker was credited as building two EVs. The extra credits could then be used to offset higher emissions from cars that didn't meet the standard.

As mentioned earlier, the GHG rule also has a 2018 checkpoint built in as a midterm review, offering a chance see whether the targets should be strengthened or relaxed. We never could have convinced the car companies to sign on without this interim review, but with 25 percent of the 2012 cars already meeting the 2016 standards and auto manufacturers investing heavily in new technologies, I want to be optimistic that the long-term standards will stick.

Yet, being optimistic about the standards in place is not enough. The EPA has to use the 2018 checkpoint to put forth the draft regulatory framework for greenhouse gas standards spanning 2026 to 2035, consistent with the practice we established for long-term, performance-based stretch standards. These standards should continue the 5 percent annual improvements under way. Continuity and consistency in the ambitious setting of standards are required if we hope to lead the global development of clean, smart cars of the future in time to head off global calamities.

## 2. Build a strong, multidisciplinary team.

Among my top priorities as the new head of the Office of Transportation and Air Quality was to build the most capable team possible—one with deep and multidisciplinary experience, outstanding skills and unrivaled expertise. Fortunately, the Ann Arbor lab already offered me a depth in technical capacity and analytical skills that matched the expertise of the auto and petroleum industries in just about everything related to emissions and fuel consumption. This was augmented by my team in DC, who were experts in the Clean Air Act law and in policy development.

Our work intersected with many industries, so we also needed to be able to tap specialized knowledge in completely different fields. Whether we were in discussions with the oil industry trying minimize capital investments in refineries, the auto and truck manufacturers trying to avoid new investment in research and development and production, or suppliers who might have to tool up for new technologies, we had to be well prepared technically and legally. If we didn't know what we were talking about, the industry representatives would have the upper hand in convincing the White House and Congress that their position was right. Just as bad, EPA leadership would lose confidence in our abilities.

Technical expertise has also proven irreplaceable outside of negotiations with regulated stakeholders. Quickly tapping our deep knowledge was essential to success in discussions with our sister federal agencies as well as Congress. For example, in 2005, Missouri senator Kit Bond threatened to hold back the EPA's budget after the lawnmower industry convinced him that the EPA-required catalysts on lawnmowers would catch fire. It was the same thing the car companies had said about the catalytic converter in the early 1970s—and just as false. Fortunately, the Ann Arbor lab was able to quickly, calmly, and irrefutably demonstrate to Senator Bond and the Association of Fire Marshalls that you can build a clean non–flammable lawn mower. Political firestorm averted, we could move forward with our work.

In 1998, when we were developing the Tier 2 clean car clean and gasoline program, we proposed a program to bring the emission requirements of SUVs to the same level as cars. The car companies initially reacted negatively. But when Chet France's engineers retrofitted a Silverado with a catalyst that met the Tier 2 standards four years earlier than was required, the car companies quietly accepted the proposal.

As I mentioned before, one measure of a successful regulatory collaboration is not getting sued afterwards. Nearly every environmental regulation is legally challenged in some way or another, but this was a rare experience for my office. We also won virtually all of those rare cases in which stakeholders did sue us. Regulations that will survive a barrage of litigation depend on compelling technical and legal arguments—but effective collaboration helps us avoid getting sued in the first place.

When the industry challenged the EPA in court on our clean diesel truck rule as not feasible, the judge pointed to the demonstration work that the Ann Arbor engineers had done as a basis for concluding that the regulations were technically sound. When we were negotiating the 2012–2016 and later the 2017–2025 greenhouse gas standards, we used the Ann Arbor lab's expertise to develop analytical models for each car company at a specific vehicle model level, showing them how they could individually meet our proposed standards using a variety of proven technologies, starting with their current fleet as a baseline. Our ability to provide specialized, authoritative data on technical and economic issues was invaluable in showing the stakeholders that we knew they could make the technical improvements we were requiring within a reasonable timeline and cost. When the White House economists questioned our technical analysis and costs for future greenhouse gas standards, we were able to provide answers based on solid scientific data and analysis. It was the strength of our multidisciplinary team that got us through the successful negotiations and gave the president and the country a historic win.

## 3. Maintain priorities, timing, and a consistent tempo.

Over the last twenty years, my team was able to put out a number of successful rules in part because we didn't try to do too much at once. There are a lot of sources of air pollution, and there were plenty of external demands on our resources; we could have rushed around giving attention to a limitless number of issues. Instead, we had a steady, continuous cadence of successful rulemaking. My office set priorities based on the highest need first, even if that meant missing statutory requirements.

For example, once we cleaned up cars and light duty tucks, then we approached heavy duty trucks, followed by non-road vehicles, locomotives, marine engines, and so on. Cleaning up construction machinery is a great accomplishment, but it didn't make sense to pursue that until after we had dealt with the much larger health problem of cars. Instead of feeding a sort of conveyor belt of regulation, we maintained our vision of the future and got the maximum benefits out of the rules we developed.

Timing is also an important part of the rhythm, and critical to political leaders' support. The first hundred days of a new administration are frequently times when the executive branch has the most political capital to spend. We were able to roll out the first greenhouse gas regulation in the window right after President Obama took office in 2009, and before his presidency became bogged down in constant fights with Congress. Eight years earlier, Christine Todd Whitman's help in clearing the diesel truck rule in her first thirty days in office during the early days of the Bush administration was equally critical. Because the Bush administration had not yet established its processes, we were able to complete the clearing of the rule quickly.

We also need to prioritize the sources of greenhouse gases. After dealing with cars and trucks, the EPA should next move to aircraft. For this, the agency will need the support of both the Federal Aviation Administration and the airlines.

Following aircraft, policymakers should look at the next phase of regulations for locomotive and marine engines. Though the EPA already has the expertise to write new rules in these sectors, the agency could initially use a strategy we developed in 2001. Smart Way was a voluntary program to improve fuel efficiency and reduce air pollutants for tractor trailers. Even though there was no real enforcement mechanism, the program became popular within the trucking industry because the compliance costs were reasonable. Likewise, the customers of the freight companies—enormous corporations like IKEA, Walmart, and Pepsi—appreciated the huge savings produced by greater fuel efficiency. Smart Way became the blueprint for our first greenhouse gas rule for commercial trucks in 2010. The agency should develop a voluntary program like Smart Way for non-road equipment and engines used in agriculture and construction as well as locomotives and marine engines as a first step in addressing greenhouse gas emissions in those sectors.

## 4. Rely on fact-based, driven collaboration.

Long before a rulemaking process begins, my office made a point of always openly sharing relevant data regarding a new rule. We weren't throwing the information over the wall but establishing transparency and the importance of technical, scientific, economic, legal, and other facts early on in the discussion.

When possible, my senior team and I also visited all the stakeholders, let them know which targets we were considering, and asked for their input. We traveled not only to Detroit but to Japan and Germany and other European and Asian countries to consult with local automakers and suppliers. We met with the oil companies, ethanol producers, states, unions, and environmental and consumer groups across the country. Though they may not have been supportive of the end goals, everyone appreciated that this was our office's standard operating procedure, and these trips and industry contact also gave us the knowledge needed to write smart regulations. During these visits, we met with presidents and senior management of the company. Our frank discussions, all conducted in confidence, were essential for building the shared trust to move forward. They knew where we stood, and we took the time to understand the challenges they would face.

Once the rulemaking clock started, however, the collaboration process was actively driven by my office. Given all their divergent viewpoints, bringing all the stakeholders into one room to reach consensus would be nearly suicidal. Likewise we avoided dealing with industry groups, which invariably negotiate from a lowest common denominator position. Instead, we relied on shuttle diplomacy and building alliances between individual stakeholders to move forward. These relationships were particularly important when a forward-thinking companies—like Ford for Tier 2 and the greenhouse gas standards; BMW, Nissan, and Hyundai for greenhouse gas standards; Cummins for the commercial truck greenhouse gas rule; Caterpillar for the non-road diesel rule; and BP for cleaner gasoline—took a leadership role to help develop successful, smart, and legally defensible rules. The leaders set the pace. The rest of the industry would, in time, follow.

My Ann Arbor engineers under Chet's leadership had, over the years, developed positive, trusting relationships with the technical teams for all the stakeholders. To help develop a successful rule, the regulated industries frequently have to share confidential data on their plans on, for example, what cars or technologies they are developing. They knew that the EPA lab would keep all their information fully confidential. Without institutionalizing this high standard of integrity, we couldn't get the best results possible.

Relationships with other agencies, especially the Departments of Energy, Transportation, Agriculture, and Office of Management and Budget, were also crucial. These agencies had their own expertise and agendas and were empowered to hold up the rulemaking process even after we had successfully negotiated with the regulated parties. Thankfully, we often had very collaborative working relationships with other agencies. With the Department of Energy, we had a very strong working relationship with Pat Davis, the Director of the Vehicle Technologies Office, and his team, as well as with the department's national labs on a number of issues from clean diesel to EVs and lightweight materials. They provided technical data that was crucial to our regulations. Two of the Department of Energy's multidisciplinary advanced energy HUB research projects, for example, are dedicated to advancing the cost efficiency of lithium-ion batteries and lightweight materials. The Department of Energy should continue these investments on battery technology, fuel cells, lightweight material and infrastructure. With cheaper, lighter, better batteries, electric vehicles will become more competitive with conventional vehicles. In the years beyond 2021, infrastructure will be even more critical.

Environmental groups almost never challenged our actions in courts. In the few cases when they did—for example, the 2007 Supreme Court case forcing EPA to take action on greenhouse gases, our legal loss felt more like a victory for science. These groups provided data, analyses, and expertise, so they too were brought in early in the process along with other stakeholders and were given the same factual data that we shared with industry, the states, and our political team.

The partnership with California is unique among regulatory agencies. California has the advantage of driving exciting initiatives without having their hands tied by the fickle politics of the federal government. Additionally, California used its unique leverage to encourage twelve other states to join it in pushing for more stringent emission standards for "California compliant" vehicles. This coalition was the springboard for the nationwide federal greenhouse gas programs we negotiated.

In the near future, I hope to see California continue to push forward national regulation, including an extension of their requirement for zero emissions vehicles to 2035 well before the current 2025 rule expires. Like most leaders, California hasn't succeeded in all of its experiments, but without the state's long-term advocacy for cleaner emissions—and its residents' love for their cars—there is no way the United States would have such progressive rules in those areas.

Though we have a special history with California, it is far from the only state we've worked with. Vermont, Massachusetts, New York, Maryland, Georgia, and Washington have been among our other most important partners. Each one had different needs. For example, Washington had an important issue with the air quality in their Pacific

ports. Georgia's sprawling capital, Atlanta, had experienced a significant growth in cars and expansion of its road network that had created
serious environmental and congestion problems. We worked successfully with all of them.

Given the trends toward urbanization and megacities, the EPA
needs to begin building closer working relationships with not just
states but major cities as they deal with congestion, traffic, pollution,
and the impacts of climate change on a rapidly increasing scale.

Good regulations were not made in a single government office in
Washington. Every connection all around the country was important.
My office door was open to adversaries as well as allies, especially when
they brought the sort of data and expertise that lends itself to successful
collaboration. Mutual respect doesn't mean always agreeing with someone but the willingness to share high quality, high integrity information. That allows for outcomes that remain true to the established goals.

## 5. Strengthen our public communication in the global village.

Following the initial round of strong regulations in the 1970s that initiated huge programs dedicated to cleaning up the country's air, land,
and water, industry groups have become more focused and effective
at slowing and weakening environmental regulatory actions. These
groups use sophisticated, well-funded, and sometimes cynical or dis-
honest lobbying efforts and front groups. Over the past few decades,
for example, the petroleum industry has led the efforts attempting to
discredit climate change as a serious threat. The oil companies' efforts
have largely relied on techniques they learned from the tobacco industry. Just as tobacco companies repeatedly obfuscated the science and
questioned the severity of smoking's negative health effects beginning
in the late 1960s, petroleum companies and some utilities have often
attempted to scare the public with well-funded studies claiming that
EPA efforts to clean up the air, water, and ground will be too costly
and cause jobs to be sent overseas.

Part of the reason industry has been able to successfully manipulate public discussion of a scientifically supported global phenomenon is that the EPA is subject to the politics of the White House and its posture on the environment. The agency's public communication efforts have traditionally consisted largely of press releases covering a new major regulatory effort. The accomplishment may be noted in the media for a couple of days before disappearing. Every day after it is implemented, that rule is making the country cleaner and people healthier, but nobody is making the connection because there is no method for regularly updating the public on the progress—or, as I like to call it, the rest of the story.

At the same time, when industry wants to comment on an EPA action, they make a big splash and engage in a public dialogue with the press and other media outlets. Though the industry's talking points are biased against agency actions—and frequently misleading or inaccurate—they remain very powerful because of the private money behind their message. These groups can afford to repeat the same misinformation over and over, sometimes for years, until they create their own reality.

Take one often-heard accusation: the EPA kills jobs. Conservative lobbies throw this claim at regulation after regulation, with no agency rebuttal. Instead, the EPA should respond publicly that many environmental rules actually result in increased employment in the United States. For example, a United States–based lithium-ion battery plant that builds $10,000 battery packs provides Americans with jobs while displacing imported oil. Nissan's Leaf battery plant in Tennessee helped create 1,300 jobs. Tesla builds its cars in an old, formerly shuttered GM plant in Freemont, California, with more than 6,000 employees and plans to build a $5 billion battery plant in Nevada that could be the largest in the world and will employ another 6,000 people. GM makes its electric motors in an American factory outside Baltimore. The BMW i series carbon fiber is made in a $300 million factory in Washington state. The Blue-Green Alliance, which represents fifteen of the nation's largest unions and environmental organizations, estimated that when vehicles achieve 40 mpg in 2020, the GHG standards will have helped add 150,000 manufacturing jobs in total.

According to a March 2011 study by the Manufacturers of Emission Controls Association (MECA), the total economic activity generated by emissions control technologies for cars and trucks adds up to $12 billion in the United States, and MECA member companies employ some 12,000 workers. Globally, the manufacture of emissions control technology is estimated to generate $36 to $43 billion in economic activity.[4] In the future, the whole American auto industry will be even more competitive globally by leading the charge in innovation to conform with the 54.5 greenhouse gas and CAFE standards. Industry will always want to avoid and discredit regulation. But when their arguments are not fact-based, they need to be actively challenged.

One of the specific difficulties in confronting climate change is that it is a global phenomenon that you can't smell, taste, or see. Asian megacities like Beijing are finally taking meaningful steps to rein in their deadly pollution, but it is also easier to stir up political action for such palpable threats. Everyone knows something is dreadfully wrong when your throat hurts from breathing and you can't see a huge building four blocks away.

In the United States, Europe, and Japan, citizens demanded that their governments confront similar problems decades ago. Today, our rivers no longer catch fire and our air isn't filled with lead exhaust or other hazardous pollutants. Proud as we are of those accomplishments, dealing with the less visible threat of climate change is more complex but just as urgent. There are plenty of data points showing that climate change is a critical problem, and most people want to curb its impact. But, because it is harder to experience first-hand, the political outcry about a dirty environment is not as passionate and immediate among many sections of the population as it was four decades ago for the much more visible conventional pollution.

What's more, climate change is not a local problem. In the 1960s, Los Angeles was choked with smog because so many people drove cars there. Acid rain fell in the industrial areas of the Great Lakes in the 1970s largely because of the dirty emissions from industries located there. Within a decade of those local sources of pollutants being cleaned up, the problems were dramatically mitigated. This is not the case with climate change, a global phenomenon in which carbon emissions in Canada help create conditions for wildfires in Australia.

Even if the direct effects of climate change are hard to see, clearly publicizing the need for continued change will build public support for policy that further reduces greenhouse gas emissions. The responsibility doesn't lie only with president or the administrator of the EPA; it benefits each one of us to be informed and take action. From individual citizens, to governments, educators, religious leaders, NGOs, the private sector, and media, we need to take action.

TV shows like the recent Showtime series *Years of Living Dangerously* serve the public well, but we need the mainstream media to concentrate on this issue and educate the public—not only when there is a rise in severe weather events but consistently. The public also needs to be better informed about the solutions available, like electric vehicles.

Most consumers do not understand these new vehicle technologies. I am continually asked how my Chevy Volt works or why I am putting gas into what is supposed to be an electric car. People must be better informed about the important contribution of EVs in fighting climate change and that the total cost of ownership is quickly approaching that of internal combustion–powered vehicles. We have to appreciate that these are new technologies that are subject to our technology adoption rate. Though our time to adopt technologies is clearly accelerating—it took only seven years for more than 25 percent of Americans to become web users versus forty-six years for electricity and twenty-six years for the TV—adoption of EVs or other clean vehicles will take time.[5] The automakers, dealers, federal, state, and local governments, and NGOs can play a role in helping consumers better appreciate their value and expand the EV ownership beyond the well informed, early adopters.

## 6. Collaborate globally.

There used to be a clear pattern in efforts to clean up air pollution from automobiles, trucks, and other transportation sources. The United States EPA, following the Clean Air Act, mandated environmental regulations that were "technology-forcing," which in turn led to innovation to reduce environmental impacts. Eventually, Europe and Japan would follow. Then the technology spread to the rest of the world. In 1970, no cars had catalytic converters. Today, only a very few cars on the planet don't have catalytic converters.

Climate change is different. While the United States clearly needs to play a leading role, greenhouse gas emissions are a kind of pollution with no boundaries. Since global warming is an urgent, worldwide phenomenon, the EPA cannot take strong measures and then wait for the rest of the world to come along. By the same token, the United States can't wait for others, like China, to take action either. In working on our climate regulations, it has become clear to me that we need to increase our capacity to support collaborative, fact-based actions globally.

The timeline for serious greenhouse gas reductions is very short. If we magically stopped all transportation-related emissions tomorrow, the air would smell cleaner tomorrow and the levels of, say, soot would begin dropping almost immediately. But the greenhouse gases from the 1990s will remain in the atmosphere for centuries. Combined with the amount of lead time it takes for a less efficient fleet to turn over—i.e. some people in 2050 will probably still be driving the more polluting cars from the late 2030s—the need for immediate action is absolute.

EPA requirements for cleaner vehicles in the United States must continue to lead innovation in advanced low carbon technologies, but we need to work more closely with China and others to have an impact globally. Although the American auto market will be smaller in comparison to China and other rapidly growing parts of the world, the United States has the know-how and ability to lead.

In the past, my office led the efforts to establish the UN process for global harmonization of transportation emissions standards worldwide.

We helped China design policies and programs to address pollution from cars and fuels and worked with numerous other governments for a sustainable transportation sector. We collaborated closely with our neighbors Canada and Mexico, who have adopted US policies largely intact. We also worked productively with the EU and Japan.

With regard to climate change, though, these efforts need to be enhanced and strengthened. While the goal of creating identical regulations worldwide presents a number of obstacles, there remain many areas where we can collaborate, including modeling, engineering advancements, scientific research, and costs data. Working toward harmonizing our testing procedures for setting greenhouse gas standards will have two major benefits. For one, global standards reduce the amount of resources each individual country must rely on to develop its own domestic regulations. Regulated companies will also save the costs associated with meeting a slew of different national testing procedures and standards, ultimately rewarding the consumer with lower cost solutions.

## 7. Enforce the regulations.

When I talk about my team's accomplishments, I tend to focus on all the work they put into writing strong, effective rules. But without consistent enforcement mechanisms, their work is useless. Simply put, the economic and public benefits to be derived from each of the regulatory efforts are only as real as the regulations' implementation and enforcement. But enforcement is not so simple. We have repeatedly witnessed public goods and human lives lost or left in turmoil because industry failed to implement rules. The results—mining accidents, oil spills, banks failing—often end up on the front page.

Ultimately, enforcement isn't about penalizing an industry but creating a level playing field for all companies. Fair competition occurs only when all players are held to the same standards. Companies not playing by the rules should face civil and, at times, criminal penalties.

In 1998, nine diesel engine manufacturers were ordered to pay the highest environmental enforcement penalty ever—$1 billion—for cheating on their emission standards test. During lab testing, the companies used the mandated emission control strategies. Once on the road, however, these emission control system were turned off. As a result, over one million tons of illegal nitrogen oxide emissions were released into the atmosphere over the course of a decade. This case highlighted to me the great importance of timely enforcement of our regulations. It took us too long to uncover the facts and act.

More recently, GM CEO Mary Barra appeared before Congress multiple times to explain why her company had for years ignored a defective ignition switch in several smaller cars. The company had argued that it did not know what had caused accidents even after internal testing had pinpointed the ignition switch problem. At least thirty deaths and thirty-one serious injuries had been linked to the defective switch as of October 2014. The company set aside $1.7 billion dollars to pay for the repairs of the 2.6 million now recalled cars and dismissed fifteen of the senior GM employees responsible for the product liability cases.

Ironically, the same Congress that held hearings harshly questioning Barra has often politicized government funding, including safety and environmental regulations. NHTSA, which has the authority to enforce automobile safety, is underfunded, making it difficult to carry out an effective safety enforcement program. With better funding for enforcement, some of the thirteen people killed in the GM accidents might not have died. Congress has attempted to weaken EPA's enforcement capacities as well. For example, some legislators tried to strip EPA budgets of funding to implement the 2016 greenhouse gas rules.

For our part, EPA has revised our regulations to make it harder for the companies to cheat, in part by implementing commonsense changes. In the case of the truck violations, for instance, we now test trucks that are actually on the road—not just the ones provided for lab testing—making for more timely and accurate assessments of compliance.

In 2010, soon after we finished the first greenhouse gas standards, I allocated additional resources to make sure car companies complied with the new standards and accurately reported fuel economy on the window stickers for new cars. I asked our compliance team to test representative car models. The result of these tests culminated in a $300-million-dollar fine on Hyundai and Kia in October 2014, the first greenhouse gas standards noncompliance penalty. This type of action assures a level playing field and serves as a deterrent for other companies.

Enforcing rules is a global issue. When I travel to China, American companies approach me with concerns about Chinese companies environmental noncompliance. China is an enormous country without a developed history of enforcing environmental regulations. But, as one of the world's largest polluters, it is also essential that they do so. EPA has worked closely with our colleagues at the Chinese EPA to share policy and regulatory approaches to implementing the Clean Air Act. And we will need to continue to do so. Strong enforcement of environmental rules is probably the biggest hurdle for China as it seeks to attack its serious environmental problems, as it is for many other nations.

I believe we need to strengthen our commitment to enforcement and bring our global partners along with us. As Tana Utley from Caterpillar pointed out to me, "Enforcement is sometimes lacking in developing countries, which makes it difficult for companies with high integrity to compete." The EPA can show leadership in ensuring the design of meaningful enforcement programs and securing funding appropriate for their implementation within our own country. We can then share these programs with others.

★   ★   ★

Six years ago, companies representing 80 percent of the US market signed on to fuel economy goals that seemed nearly impossible in the 1990s, a time when GM was repossessing and destroying their electric cars. Impressively, automakers are now succeeding at meeting these goals while remaining profitable. Many manufacturers are now introducing vehicles years ahead of the target dates. Almost 28 percent of model year 2013 vehicle sales met the greenhouse gas and mileage targets for 2016. Five percent of 2013 vehicles already meet the 2025 standard. After decades of undermining and resisting increases in fuel economy—while ignoring greenhouse gas emissions—automakers are now competing to develop the smartest and most cost effective technologies to reduce carbon.

Tom Linebarger, CEO of Cummins, one of the most forward-thinking companies on environmental issues, articulated the role of regulations in innovation in a conversation in October 2013. He rejects the notion that innovation happens only when the governments stay out of the way of an industry. Instead, he believes investments in cleaner engines and more efficient trucks—which save the truck fleet operators significant fuel costs—would not happen without government regulations. He went on say that "meeting EPA emissions standards in the US has helped us innovate and be ahead of the competition, and grow overseas markets." For example, Cummins's updated, cleaner diesel engine introduced in China in 2013 cost $255 less than what it was first introduced in the US over ten years ago.

In 2018, I fully expect that the car companies' success will play a big role not only in affirming the current 2025 standards but in setting further goals for 2035 and beyond. I expect to see the world's largest automakers continuing to invest in advanced technologies like lower cost batteries, higher performing hybrids, connected or driver-less vehicles, and other kinds of personal transportation that seemed like science fiction just a few years ago. The future comes faster every year and, with it, the race to reduce the amount of greenhouse gas building up around our planet.

The battle to achieve reductions in greenhouse gases became the toughest challenge of my career but also yielded my proudest accomplishment. I remain convinced that, by 2050, we can leave the world in better shape than we found it. For our children and grandchildren to avoid the worst of climate change, we have no other choice.

# AFTERWORD

There was not one day that I did not treasure the job I had in the EPA, working to save lives and protect the world around us. The job wasn't always easy. Sometimes we were ridiculed. One Congressman publicly referred to me as a "faceless bureaucrat" who was destroying farming, while another publicly called me a liar with no justification. We were trashed in the press and the media. The criticisms always outweighed the compliments.

Yet my team and I believed in our mission and worked as hard as we could to make a measurable impact on the lives of the citizens we were tasked with protecting. Clean air is beneficial to everyone regardless of where they live, their income level, race, gender, politics, or religion. We wanted to make sure that the government was working to everyone's benefit. That is all that mattered. We believe we succeeded and can only hope that those who follow us do more and do it better.

Future efforts will be even more important as we face the most extraordinary environmental challenge of climate change. Our inaction will have devastating impacts on every corner of the planet and on each one of us, especially future generations. Climate change threatens our food supply, water sources, places where we live and work, our national security, and our economy. We have a moral responsibility to act now, and history will judge us by how we as people, and governments, rise to this challenge.

# ACKNOWLEDGMENTS

I have been immensely privileged ever since I arrived to the United States nearly fifty years ago. Though I had almost no knowledge of the English language, I was soon able to study engineering at Lowell Technological Institute (now the University of Lowell) with some of the greatest people in this country. Later I was also honored to work at the Environmental Protection Agency for thirty-two years. Every day as a public servant, I knew I was making Americans' lives healthier and more productive. Even after years at the agency, I was constantly amazed at how much we could accomplish—despite the politics and resistance from industry and, at times, from the White House and Congress. After I retired from EPA, I set out to tell the story of the many unsung heroes in federal and state government as well as nonprofit and philanthropic organizations who were critical in establishing the first regulatory program that reduced greenhouse gas emissions from cars. This same program has also paved the road for the car of the future.

I want to recognize Cal Barksdale, executive editor of the Arcade Publishing imprint at Skyhorse Publishing. When others gave up on my story, not only did he believe in it but helped me write a stronger book by shaping the content and doing a wonderful job editing the manuscript. Special thanks go to Marc Adelman of Adelmania Consulting who supported this project from day one and was invaluable in guiding me through the unfamiliar publishing industry. I also thank my literary agent Carol Mann as well as Steve Perrin of Galvanized.

There are two individuals without whom it would have been impossible to write this book. My wonderful collaborator, Nathan Means, consistently brought creativity, dedication, and passion to this project. He also helped with the research, creatively transformed my writings, and provided tremendous support throughout. My husband, Cuneyt, an automotive industry expert, advised me and worked with me on a daily basis. On days I was discouraged, he believed in me and helped me see the project through. I could never have completed this book without his encouragement and support. Allison Fitzgerald helped me with the first book proposal, and Ryan Pretzer and Adam Ahmad helped with the research for the book. John Koupal of ERG Inc. generously helped with the modeling. The book also benefited from interviews of many people I have worked with from government, industry, the states, and various organizations. I thank them for their time and insights.

As I set out to write the book, I received the generous support of Charlotte Pera of Climate Works as well as Jason Mark and Patty Monahan of the Energy Foundation. These groups, along with the Hewlett Packard Foundation, have consistently supported strong and smart policies in the transportation sector.

From the beginning of this project, I also received critical assistance from Drew Kodjak of the International Council for Clean Transportation, a group that has very successfully worked with many government agencies and other stakeholders across the planet to promote clean transportation.

Despite how fortunate I have been in my career, I am still surprised at how much effort our society spends on bashing our government—and how infrequently our dedicated public employees are recognized for their many positive contributions. There are countless unsung government employees who miss holidays and vacations and go beyond the call of duty to carry out their responsibilities to improve and protect the Americans' air, water, and overall environment.

I know this because, during my time at the EPA, I was fortunate enough to work with far too many extremely talented and dedicated people to mention all by name. However, I must acknowledge a few of the most indispensable members of my staff and other partners. First I want to mention my Ann Arbor team, beginning with my right-hand man, Chet France, who had impressive technical expertise, passion, and high standards for his team's work. He never complained, even the countless times when I made unreasonable requests of him to get things done to meet impossible milestones. Christopher Grundler, my deputy, supported our efforts for eighteen years and made sure that we had all the capabilities and the talent needed to do the work. The lab also offered the advice of experts like Bill Charmley, a manager who oversaw all the technical analyses for our rules. Dr. Charles Gray, one of the Ann Arbor lab's first engineers, has been an invaluable EPA contributor for years. Another key contributor, Byron Bunker, had previously worked as an engineer for Cummins and knows heavy-duty diesel engines inside out. Paul Machiele was an expert on refinery issues who argued successfully with the oil industry on new fuel standards.

Karl Simon, in our Washington, DC, office, trained as a lawyer and served in many different capacities. His expertise in dealing with the Office of Management and Budget and assisting with international dialogue proved invaluable. Sarah Dunham directed our climate policy and renewable fuels efforts, with an eye to recruiting great staff and working on complex policy issues.

But the technical knowledge was just one part of the battle. On the policy side, I had one of the agency's best attorneys, John Hannon, who, along with his team, which included two other brilliant lawyers, Steve Silverman and Michael Horowitz, helped me write the most competent and airtight rules that the EPA had ever seen.

Maureen Delany, a special policy advisor, could both understand the big picture and work the details of complex legal and technical issues. She also brought along a good sense of humor that we desperately needed at times. I also have to thank my chief of staff, Karen Orehowsky, for whom no job was too small or big, and honor the memory of the late Gladys Bryant, my tireless assistant for 23 years at the EPA.

Over the years at EPA, I've had many excellent bosses I would also like to recognize. In 1994, it was Mary Nichols, then the assistant administrator of the Air and Radiation Office of the EPA and now the chairwoman of California's Air Resources Board, who appointed me as the director of the Transportation and Air Quality office. Throughout the years, she has been a strong voice and a force promoting clean transportation. Bill Reilly, EPA administrator under President George W. Bush, was among my first heroes and a long-standing mentor. Carol Browner, EPA administrator under president Clinton, and Assistant Administrator for Air Bob Perciasepe provided the leadership for the clean car/fuels and diesel truck rules. Christine Todd Whitman, EPA administrator under President Bush, helped save the diesel truck rule from being squashed by the administration. Lisa Jackson, EPA administrator under President Obama, Gina McCarthy, now EPA administrator, and Lisa Heinzerling, climate counsel to the administrator, were the three women who brought the courageous leadership necessary to see through the first US climate actions.

Collaboration with Mary Nichol's deputy at the California Air Resources Board, Tom Cackette, and their staff was also key to our success. Working with California, sharing technical data and models and policy ideas, and collaborating on all aspects of the transportation and fuel regulations made both our teams stronger in our effort to work with the regulated industries and develop smart rules and policies.

In addition to collaborating with California on all of our major programs, we formed a strategic and mutually supportive working relationship with Bill Becker, the executive director of the National Association of Clean Air Agencies (NACAA). The group represents forty-five states and territories across the United States. For over thirty years, Bill has been a vocal champion of clean air. Representing NACAA,

Bill provided important information about his member states and supported the federal actions to clean the air from transportation.

Mike Walsh, the founder of the International Council for Clean Transportation, not only has spent over three decades helping countries and governments take actions toward clean transportation but has been a great supporter and friend. I also offer thanks to Mike who, along with John German, my longtime friend Ellen Berick, Maureen Delany, and John Hannon, reviewed the manuscript and offered thoughtful input.

There are so many great staff at numerous environmental and public health groups who have worked endlessly supporting environmentally smart actions. To mention just a few of these many groups and individuals, thanks to Michelle Robinson and David Friedman with the Union of Concerned Scientists; Rich Kassel, David Doniger, Rich Kassel, and Ronald Hwang with the National Resources Defense Council; Vickie Patton with the Environmental Defense Fund; and Paul Billings with the American Lung Association. Although these groups don't have the industry's deep pockets, they produced strong technical analysis on the need, cost, and legal justification of actions taken by my office. Very often they were further ahead than we were, pushing for more stringent regulations. We could almost always count on them to support our efforts even when the results were not exactly what they had been advocating.

I also want to thank many colleagues across the industries we regulated who treated me and my staff with respect, regardless of whether we agreed or not. Many companies showed leadership in helping craft effective environmental policies and worked collaboratively with my office. And I would like to thank the engineers in the transportation industry, and especially the emissions controls manufacturers, for their innovations that have helped reduce the conventional pollution from tailpipes to levels we could have never imagined. Their ingenuity and dedication have helped save lives and improve the health of people worldwide. Special thanks to all those companies developing the cleaner, smarter car technologies of the future.

Finally, I want to thank my two daughters, Nicole and Marisa, who endured my long hours at the EPA and supported me as I missed soccer games, PTA meetings, and vacations. I thank both my girls for their help with the book and, for their creative design for the book's cover, my son-in-law Enrique and Marisa. I do everything I can to leave this planet a better, cleaner place than I found it. It's because of them, their children, and all our children and grandchildren that we must succeed in our efforts to address the catastrophic potential impacts of climate change.

# ABBREVIATIONS

BEV: battery electric vehicle, e.g. Tesla S, Nissan Leaf
CAA: Clean Air Act
CAFE: Corporate Average Fuel Economy
CARB: California Air Resources Board
CNG: compressed natural gas
$CO_2$: carbon dioxide, a greenhouse gas
DOE: Department of Energy
DOT: Department of Transportation
EDF: Environmental Defense Fund
EIA: Energy Information Agency
EPA: Environmental Protection Agency
FEV/FCEV: fuel cell vehicle
ggt: gigatons (1000 tons)
GHG: greenhouse gases
HEV: hybrid electric vehicle; e.g. Toyota Prius
ICCT: International Council on Clean Transportation
ICE: internal combustion engine
IEA: International Energy Agency
IPCC: Intergovernmental Panel on Climate Change
kwh: Kilowatt-hours is a measure of the electrical energy capacity of a battery pack for an Electric Vehicle.

MECA: Manufacturers of Emission Controls Association

MMTCDE: Million metric tons of carbon dioxide equivalent

mpg: miles per gallon (of gasoline or diesel)

mpge: miles per gallon "equivalent"; used to compare energy consumption of alternative fuel vehicles, plug-in electric vehicles, and other advanced technology vehicles with the fuel economy of conventional internal combustion vehicles, expressed as miles per US gallon.

MY: model year, used to describe the year of the vehicle model e.g. MY 2015

NACAA: National Association of Clean Air Agencies

NHTSA: National Highway Transportation Safety Agency

NRDC: National Resources Defense Council

NOX: nitrogen oxide, a conventional pollutant

OECD: Organization for Economic Cooperation and Development (includes fourteen developed countries)

OMB: Office of Management and Budget

OPEC: Organization of Petroleum Exporting Countries

PHEV: plug-in electric vehicle; e.g. Chevrolet Volt

pm: particulate matter, a conventional pollutant

ppm: parts per million; number of parts of a chemical found in one million parts of a particular gas, liquid, or solid.

# NOTES

## Introduction

1. www.americanprogress.org/issues/green/news/2013/02/12/52881/going-to-extremes-the-188-billion-price-tag-from-climate-related-extreme-weather/.

## Chapter 1

1. "Global Energy Perspectives with Focus on Sustainable Energy for Green Economic Growth," Nebojša Nakićenović, DTU International Energy Conference, Sustainable Energy for Green Economic Growth, Copenhagen, September 10–12, 2013.
2. www.epa.gov/40th/achieve.html.

## Chapter 2

1. http://cnnpressroom.blogs.cnn.com/2014/06/08/president-of-kiribati-anote-tong-on-climate-change-its-too-late-for-us-on-cnns-fareed-zakaria-gps/.
2. Robin McKie, "Miami, the great world city, is drowning while the powers that be look away," *Guardian,* July 11, 2014.
3. Ibid.
4. www.scientificamerican.com/article/massive-seawall-may-be-needed-to-keep-new-york-city-dry/.

5.  Intergovernmental Panel on Climate Change, *Climate Change 2013: The Physical Science Basis*, http://www.climatechange2013.org/.

6.  riskybusiness.org/uploads/files/RiskyBusiness_Report_WEB_09_08_14.pdf.

7.  "Climate Change Impacts in the United States: Highlights," US Global Change Research Program, 2014.

8.  www.epa.gov/climatechange/science/causes.html.

9.  E. P. Evans: "The Authorship of the Glacial Theory," *North American Review* 145, no. 368 (July 1887).

10. John Tyndall, *Contributions to Molecular Physics in the Domain of Radiant Heat: A Series of Memoirs Published in the 'Philosophical Transactions' and 'Philosophical Magazine' with Addition* (New York: D. Appleton and Company), 424.

11. John Gribbon and Mary Gribbon, *James Lovelock: In search of Gaia* (Princeton, NJ: Princeton University Press, 2009).

12. Svante Arrhenius, *Worlds in the Making: The Evolution of the Universe*, trans. by H. Borns (New York: Harper and Brothers, 1908), 63.

13. "The Discovery of Global Warming," American Institute of Physics, www.aip.org/history/climate/Revelle.htm.

14. Roger Revelle and Hans Suess, "Carbon Dioxide Exchange Between Atmosphere and Ocean and the Question of an Increase of Atmospheric $CO_2$ during the Past Decades," Scripps Institution for Oceanography, University of California, La Jolla, CA, 1957.

15. Mayer Hillman, Tina Fawcett, and Sudhir Chella Rajan, *The Suicidal Planet* (New York: Thomas Dunne Books, 2007).

16. Walter Sullivan, "Scientists Fear Heavy Use of Coal May Bring Adverse Shift in Climate," *New York Times*, July 25, 1977, A1.

17. Michael D. Lemonick, "The Heat Is On: How the Earth's Climate Is Changing," *Time*, October 19, 1987.

18. Richard J. Bord, Ann Fisher, and Robert E. O'Connor, "Public Perceptions of Global Warming: United States and International Perspectives," *Climate Research* (Pennsylvania State University) 11 (December 17, 1998): 75–84.

19. http://cdiac.ornl.gov/trends/co2/siple.html.

20. http://climatechange.umaine.edu/icecores/IceCore/Home.html.

21. www.acecrc.org.au/Research/Southern%20Ocean%20Carbon%20Sink.

22. http://climatechange.umaine.edu/icecores/IceCore/Home.html.

23. Lewis A. Owen, Robert C. Finkel, Richard A. Minnich, and Anne E. Perez, "Extreme Southwestern Margin of Late Quaternary Glaciation in North America: Timing and Controls," Geological Society of America, *Geology*, August 2003.

24. "Atmosperic Concentration of Greenhouse Gases," Environmental Protection Agency, http://cfpub.epa.gov/eroe/index.cfm?fuseaction=detail.viewPDF&ch=46&lShowInd=0&subtop=342&lv=list.listBy Chapter&r=239797.

25. "IPCC Fourth Assessment Report, Summary for Policymakers," www. ipcc.ch/pdf/assessment-report/ar4/wg1/ar4-wg1-spm.pdf, 5.

26. William Robbins, "Drought-stricken Areas Find Relief after Rains," *New York Times*, September 16, 1989.

27. Philip Shabecoff, "Global Warming Has Begun, Expert Tells Senate," New York Times, June 24, 1988.

28. www.csmonitor.com/USA/2011/0711/As-much-of-US-swelters-here -are-5-worst-heat-waves-of-past-30-years/Summer-1988.

## Chapter 3

1. www.margaretthatcher.org/document/107817.

2. Roger A. Pielke Jr., "Policy History of the US Global Change Research Program: Part I. Administrative Development," September 1999.

3. "Administration Split on Warming Issue, Aides to Bush Say," *New York Times*, February 3, 1990.

4. Philip Shabecoff, "Bush Asks Cautious Response to Threat of Global Warming," *New York Times*, February 6, 1990, Section A.

5. Stephen Engelberg, "Sununu Says He Revised Speech on Warming," *New York Times*, February 5, 1990.

6. Philip Shabecoff, "Bush Denies Putting Off Action On Averting Global Climate Shift," *New York Times*, April 19, 1990.

7. Stanton A. Glatz and John Slade, eds., *The Cigarette Papers*, (Berkley and Los Angeles: University of California Press, 1996), 189.

8. www.aip.org/history/climate/public2.htm.

9. Andrew Revkin, "Industry Ignored Its Scientists on Climate," *New York Times*, April 23, 2009.

10. George Monbiot, "The Denial Industry," *Guardian*, September 18, 2006.

11. Brenda Ekwurzel, personal interview, September 5, 2013.

12. William Clinton, "Address Before a Joint Session of Congress on Administration Goals," February 17, 1993, www.presidency.ucsb.edu/ws/?pid=47232.

13. Natalie Goldstein, *Global Warming* (New York: Infobase Publishing, 2009), 151.

14. Ibid., 153.

15. IPPC, "Climate Change 2001: Synthesis Report Summary for Policy-makers," 2.

16. September 29, 2000, announced plan, web.archive.org/web/2001011103 5000/http://www.georgebush.com/News/speeches/092900_energy.html.

17. Ron Suskind, *The Price of Loyalty* (New York: Simon and Schuster, 2004), 109.

18. www.pbs.org/wgbh/pages/frontline/hotpolitics/interviews/whitman. html.

19. Andrew Revkin, "Bush Aide Softened Greenhouse Gas Links to Global Warming," *New York Times*, June 8, 2005.

20. Andrew Revkin, "Climate Expert Says NASA Tried to Silence Him," *New York Times*, January 29, 2006.

## Chapter 4

1. Dan Becker, personal interview, July 8, 2013.

2. www.noaanews.noaa.gov/stories2009/20090113_ncdcstats.html.

3. www.prnewswire.com/news-releases/toyota-is-global-hybrid-leader-with-sales-of-7-million-279077081.html.

## Chapter 5

1. Murray Schumach, "Smog Swept Away by Cool Air Mass; Emergency Ended," *New York Times*, November 7, 1966.

2. www.epa.gov/40th/achieve.html.

3. Michael Walsh, personal interview, August 28, 2013.

4. E. W. Kenworthy, "G.M. Asks Delay on '75 Standards; Says It Can't Meet Limits on Two Auto Pollutants—Ruckelshaus Unsatisfied," *New York Times*, March 13, 1973.

5. Ibid.
6. *Toledo Blade*, September 11, 1974.
7. Michael Walsh, personal interview, August 28, 2013.

## Chapter 6

1. David Donigar, personal interview, March 14, 2013.
2. Ibid.
3. Lisa Heinzerling, personal interview, August 20, 2013.
4. John Hannon, personal interview, July 16, 2013.
5. David Donigar, personal interview, March 14, 2013.

## Chapter 7

1. "2009 Washington Auto Show," *Washington Post*, February 2, 2009.
2. Keith Bradsher, *High and Mighty: The Dangerous Rise of the SUV* (New York: Perseus Book Group, 2002), 113.
3. www.fueleconomy.gov/feg/pdfs/420r13011_EPA_LD_FE_2013_TRENDS.pdf.
4. Jamie LeReau, "LaNeve to Chevy: Fix the Car Ads," *Automotive News*, June 4, 2007.
5. http://usnews.rankingsandreviews.com/cars-trucks/daily-news/090107-The-Ten-Best-Selling-Vehicles-of-2008/.
6. The Monitor's View, "Big 3 Pump Up While Drivers Pay Out," *Christian Science Monitor*, August 15, 2014.

## Chapter 8

1. As quoted in Jack Doyle, *Taken for a Ride: Detroit's Big Three and the Politics of Pollution* (New York: Four Walls, Eight Windows), 2000.
2. Daniel Yergin, *The Quest: Energy, Security and the Remaking of the Modern World* (New York: Penguin Books, 2011), 686.
3. Fran Pavley, personal interview, October 22, 2013.
4. Fran Pavley, personal interview, October 17, 2013.
5. James Marston, personal interview, November 4, 2014.

6. John M. Broder and Felicity Barringer, "E.P.A. Says 17 States Can't Set Emission Rules," *New York Times*, December 20, 2007.

## Chapter 9

1. http://en.wikipedia.org/wiki/February_2009.
2. Ron Medford, personal interview, July 22, 2013.
3. Jason Burnett, personal interview, July 9, 2013.
4. Dan Becker, personal interview, July 8, 2013.
5. http://georgewbush-whitehouse.archives.gov/news/briefings/20010507.html.
6. Jason Burnett, personal interview, July 9, 2013.

## Chapter 10

1. Ian Talley, "EPA Set to Move toward Carbon-Dioxide Regulation," *Wall Street Journal*, February 23, 2009.
2. www.reuters.com/article/2009/02/26/us-gm-idUSN2653343220090226.

## Chapter 11

1. Chet France, personal interview, May 22, 2013.
2. Bill Charmley, personal interview, May 22, 2013.
3. "Early Resolve: Obama Stand in Auto Crisis," *New York Times*, April 28, 2009.

## Chapter 12

1. Jody Freeman, personal interview, July 11, 2013.
2. Ibid.
3. www.whitehouse.gov/the_press_office/President-Obama-Announces-National-Fuel-Efficiency-Policy/.

## Chapter 13

1. Ron Bloom, personal interview, July 11, 2013.
2. Bill Charmley, personal interview, May 22, 2013.

3. Mary Nichols, personal interview, July 9, 2013.

4. www.whitehouse.gov/blog/2012/08/28/historic-fuel-efficiency-standards-cars-and-light-trucks.

5. Bill Vlasic, "U.S. Sets Higher Fuel Efficiency Standards," *New York Times*, August 28, 2012.

## Chapter 14

1. "The Cost of Delaying Action to Stem Climate Change," July 2014, White House.

2. "Transitions to Alternative Vehicles and Fuels," National Academy of Sciences, 2013.

3. "Global Transportation Energy and Climate Roadmap," ICCT, 2012.

4. Mike Wall, "Automotive Industry Outlook: Assessing Opportunities in a Multi-Speed World," IHS Automotive, May 2013.

5. "International Energy Outlook," International Energy Agency, 2013.

6. "Cities and Climate Change," *UN-HABITAT Report*, 2011.

7. "Cities and Climate Change: Policy Directions Global Report on Human Settlements," United Nations Human Settlements Programme, 2011.

8. "Health Impacts of Road Transport," OECD, 2014.

9. Andreas Mai, "The Internet of Cars: A Business Case," Intelligent Transportation Systems World Congress, Cisco, 2011.

10. "Climate Action in Megacities," C40 Cities Baseline and Opportunities, vol. 2.0, 2014.

11. Ulrich Kranz, personal interview, October 8, 2013.

12. "US Energy Information Administration Energy Outlook," United States Energy Information Administration, 2013.

13. PWC Autofacts, 2014 Q1 Data Release.

14. "Yesterday's fuel," *Economist*, August 3, 2013.

15. James Woolsey, personal interview, October 24, 2013.

16. www.costsofwar.org/sites/default/files/Summary%20Costs%20of%20War%20NC%20JUNE%2026%202014.pdf.

17. "Oil, Economic Growth and Plug-In Electric-Hybrid Vehicles: Why Energy Is an Economic Planetary Emergency," Erice International Seminars on Planetary Emergencies, August 22, 2013.

18. Henry Kelly, "U.S. Department of Energy Vehicle Technologies Program Overview," 2011.
19. David Strickland, personal interview, August 10, 2013.
20. "Alix Partners Car Sharing Outlook," Alix Partners, February 5, 2014.

## Chapter 15

1. Rick Gore, "Ready to Race Against Smog," *Life*, September 4, 1970.
2. Automotive Engineering International, SAE International, September 3, 2013.
3. "AAA: Stop-start Engines Save Gas, Cut Emissions," *USA Today*, July 23, 2014.
4. Richard Truett, "Ford, GM Take Opposite Routes to Engine Fuel Economy," *Automotive News*, January 6, 2014.
5. "Transitions to Alternative Vehicles and Fuels," National Academy of Sciences, 2013.
6. Ulrich Kranz, personal interview, October 8, 2013.
7. "Electric Vehicle Market Forecasts: Global Forecasts for Light Duty Hybrid, Plug-In Hybrid, and Battery Electric Vehicle Sales and Vehicle Parc: 2013–2022," Navigant Research, December 2013.
8. "Transitions to Alternative Vehicles and Fuels," National Academy of Sciences, 2013.
9. www.tesla.com.
10. "Electric Vehicle Market Forecasts: Global Forecasts for Light Duty Hybrid, Plug-in Hybrid, and Battery Electric Vehicles: 2013–2020," Navigant Research, September 30, 2013.
11. www.consumerreports.org/cro/2013/03/electric-cars-101/index.htm.
12. "Global EV Outlook: Understanding the The Electric Vehicle Landscape to 2020," International Energy Agency, April 2013.
13. "Transitions to Alternative Vehicles and Fuels," National Academy of Sciences, 2013.
14. "The China New Energy Vehicles Program: Challenges and Opportunities," PRTM and World Bank, 2011.
15. "Hydrogen and Fuel Cell Technical Development and Commercialization Activity," 2013 Annual Report of the Hydrogen and Fuel Cell Technical Advisory Committee, May 2014.

16. Nichola Groom and Chnag-Ran Kim, "Toyota Aims to Replicate Prius Success with Fuel Cell Mirai," Reuters, November 18, 2014.
17. "Hydrogen and Fuel Cell Technical Development and Commercialization Activity."
18. Ibid.
19. "Toyota to Begin Selling Hydrogen Fuel Cell Car Mirai for First Time," *Guardian*, November 18, 2014.
20. "Hydrogen and Fuel Cell Technical Development and Commercialization Activity."
21. "Transitions to Alternative Vehicles and Fuels," National Academy of Sciences, 2013.

## Chapter 16

1. http://newsoffice.mit.edu/2011/cars-on-steroids-0104.
2. Venkat Srinivasan, personal interview, September 3, 2013.
3. "McKinsey Report: Lightweight, Heavy Impact," McKinsey and Company, 2012.
4. "Driving Good Dynamics," *Automotive Engineering International*, May 6, 2014.
5. Venkat Srinivasan, personal interview, September 3, 2013.
6. "Market Outlook: Surplus in Carbon Fiber's Future?," *Composites World*, March 1, 2013.
7. www.greencarcongress.com/2013/12/20131220-ford.html.
8. Andrew Brown, personal interview, June 16, 2014.

## Chapter 17

1. www.cngprices.com.
2. www.eia.gov/dnav/pet/hist/LeafHandler.ashx?n=PET&s=EMM_EPM0_PTE_NUS_DPG&f=W.
3. Coral Davenport, "Study Finds Methane Leaks Negate Benefits of Natural Gas as a Fuel for Vehicles," *New York Times*, February 13, 2014.
4. http://cities-today.com/2014/01/sao-paulo-to-introduce-its-first-fleet-of-fully-electric-buses/.

5. "Smart Bioenergy: Guiding Sustainable Bio-based Energy and Fuels Development," Union of Concerned Scientists, May 2010.

6. "Sports Cars Embrace Array of Green Technology," *Automotive Engineering International*, July 2014.


## Chapter 18

1. Kevin Ashton, "That 'Internet of Things' Thing," *RFID Journal*, July 22, 2009.

2. Joann Muller, "Silicon Valley vs. Detroit: The Battle for the Car of the Future," *Forbes Magazine*, May 27, 2013.

3. Ibid.

4. Andreas Mai, personal interview, June 16, 2014.

5. Tana Utley, personal interview, July 10, 2014.

6. Ron Medford, personal interview, June 30, 2014.

7. www.epa.gov/cleanenergy/energy-resources/refs.html.

8. Robbie Diamond and Sam Ori, personal interview, September 13, 2013.

9. "Look Ma, No Hands: Forging into a New (Driverless) World," PWC Autofacts, February 18, 2013.

10. www.seapine.com/pr.php?id=217.


## Chapter 19

1. Greg Bishop, "At the Australian Open, It's Not the Heat, It's the Stupidity," *New York Times*, January 17, 2014.

2. Lindsay Chappell, "Fuel Regs Will Stymie Industry after 2018, NADA Economist Says," *Automotive News*, August 5, 2014.

3. "Report to Congress on the Costs and Benefits of Federal Regulations and Unfunded Mandates on State, Local and Tribal Entities," Office of Management and Budget, 2001–2005 Reports.

4. "MECA Highlights Economic Benefits of Emission Control Technologies," Manufactures of Emissions Controls Association press release, March 11, 2011.

5. www.pewresearch.org/fact-tank/2014/03/14/chart-of-the-week-the-ever-accelerating-rate-of-technology-adoption/.

# INDEX